Concrete

Neville's Insights and Issues

Adam Neville

CBE, DSc(Eng), DSc, PhD, FICE, FIStructE, FREng, FRSE
Hon LLD (St Andrews and Dundee), DAppSci h.c., Hon Fellow, Queen Mary,
  University of London

Honorary Member of the American Concrete Institute
Honorary Member of the Concrete Society

Formerly
Head of Department of Civil Engineering, University of Leeds, England
Dean of Engineering, University of Calgary, Canada
Principal and Vice-Chancellor, University of Dundee, Scotland
President of the Concrete Society
Vice-President of the Royal Academy of Engineering

thomas telford

Published by Thomas Telford Publishing, Thomas Telford Ltd,
1 Heron Quay, London E14 4JD.
www.thomastelford.com

Distributors for Thomas Telford books are
USA: ASCE Press, 1801 Alexander Bell Drive, Reston, VA 20191-4400
Japan: Maruzen Co. Ltd, Book Department, 3–10 Nihonbashi 2-chome, Chuo-ku, Tokyo 103
Australia: DA Books and Journals, 648 Whitehorse Road, Mitcham 3132, Victoria

First published 2006

Also available from Thomas Telford Books
Role of Cement Science in Sustainable Development. R. K. Dhir, M. D. Newlands and L. J.
Csetenyi. ISBN: 07277 32463
Historic Concrete: The background to appraisal. Edited by J. Sutherland, D. Humm and M.
Chrimes. ISBN: 07277 2875X

A catalogue record for this book is available from the British Library

ISBN: 0 7277 3468 7

© Adam Neville 2006

Typeset by Academic + Technical, Bristol
Printed and bound in Great Britain by MPG Books, Bodmin, Cornwall

To

Mary Hallam Neville

With love and deep gratitude for her enormous contribution

and for

My children: Elizabeth Louise and Adam Andrew

And my grandchildren: Matthew, Katherine, Claire, Laura and Sophie

Acknowledgements

I am grateful to Pierre-Claude Aïtcin for his agreement to include in this book Section 2.1, which was written jointly by him and myself. Likewise, I am grateful to R.E. Tobin for his agreement to include in this book Section 4.4, written jointly by him and myself.

I owe thanks to those publishers who permitted me to reproduce in this book my original papers and articles, details of which are given in the Appendix on pp. 304–306.

These publishers are as follows:

- Precast/Prestressed Concrete Institute, for Sections 5.2 and 6.1.
- American Concrete Institute, for Sections 2.1, 4.1, 4.2 and 6.4.
- *The Structural Engineer*, for Sections 3.3, 4.5 and 6.3.
- Springer Science and Business Media, for Section 2.2.
- The Concrete Society, publisher of *Concrete*, England, for Sections 3.1, 3.2, 6.2, 6.5, 7.1, 7.2 and 7.3.
- Elsevier, publisher of *Cement and Concrete Research*, for Section 4.3.
- Concrete Home Building Council, publisher of *Residential Concrete*, for Sections 5.3 and 5.4.

Adam Neville
London, March 2006

Contents

1

Introduction

This book is a collection of papers and articles written by me in the last four years, presented as sections. In the majority of cases, the sections are exact copies of the original published material; in other cases, some changes have been made to improve or correct the text. Each chapter begins with a specially written preliminary part. Of course, the content and format are my responsibility.

About the author

To make sure that people read this book, or better still buy it, I should say, casting modesty aside, that I have written nine other books, of which seven are on concrete. One of these, *Properties of Concrete*, first published in 1964, has appeared in four editions in English and also in 12 other languages, in some cases with more than one edition translated into a given language. Over the period of 40 years, since *Properties of Concrete* first appeared in print, well over half a million copies have been sold.

On the negative side, the latest and last edition of *Properties of Concrete* appeared in 1995 in the UK, in 1996 in the USA, and 2000 in India. There have been 12 impressions of the British edition, containing some improvements, emendations and minor additions but the treatment of some very recent developments is absent.

About the approach of this book

Of course, much can be written about recent developments in concrete and also a different approach is possible. This is why, in 2003, I published a new book under the title *Neville on Concrete*, which deals with several aspects of concrete not covered, or only partially covered, previously. But that is not the main feature of *Neville on Concrete*.

Unlike usual books on concrete, which proceed in an orderly and systematic manner from science to practice, the various sections of *Neville on Concrete* first look at a problem or an issue, and then discuss the underlying scientific and technological aspects. This is like looking at concrete through the wrong end of the telescope, which gives some new insights.

I have largely followed the same approach in the present book, *Concrete: Neville's Insights and Issues*. For example, Sections 4.1 and 4.2 deal with a relation between the age of cracks in concrete and the observed depth of carbonation, a proposition advanced in an insurance case. To prove, or disprove, that relation, the relevant factors influencing the progress of carbonation are considered in detail. The proposition, which was used in litigation, is found to be erroneous. But what is important for a reader of this book is that the relevant sections in Chapter 4 take him or her through the actual pattern of carbonation and provide a scientific understanding of the phenomena involved in a manner more palatable than would be used in a classical text, which starts with the relevant chemical reactions, their kinetics, and the observed changes in the microstructure of the hydrated cement paste.

Another distinguishing feature of this book is that it has mainly been written by a sole author, so that the inconsistencies in terminology and internal contradictions are avoided. The only exceptions are Section 2.1, written jointly with P.-C. Aïtcin, and Section 4.4 co-authored by Bob Tobin. I am grateful to them for agreeing to include our joint papers in 'my' book.

I have referred to litigation. Indeed, several sections of this book have their origin in lawsuits in which I appeared as an expert witness. Now that I no longer undertake court work, I feel publicly free to discuss topics on which I provided expert opinion. Of course, I do not identify individual cases or parties but the lessons to be learned are aired.

The relevance of litigation to the structural engineer is the subject of Section 6.3. Alas, this is a topic of increasing importance because we live in a progressively more and more litigious society, and it behoves us to know the potential pitfalls. To say that litigation benefits no one would be incorrect because a large number of lawyers make good money out of it. The lawyers, who are advocates and not technical people, need technically competent experts. If, in addition, these experts can present their expertise well, if they can robustly resist occasional onslaught by lawyers for the opposing side because they know what they know and, if they neither vacillate nor exceed their field of competence – in other words, they know what they don't know – such experts can command high honoraria.

The experts' contribution to establishing technical facts is essential, but we would all do much better without litigation. A great deal of uncertainty would be removed, money expended on lawyers' fees and experts' fees would be saved, insurance premiums for professional work and for construction in general would be lower. In the end, the cost of constructing a building would be lower, which would of course benefit the owner.

This book does not deal with all aspects of concrete – indeed, it is not intended to be an encyclopedia, but rather a selected treatment of topics on which I have 'hands-on' experience. For an encyclopedic presentation, I cannot resist advertising *Properties of Concrete*.

Different people, especially in different countries, will find different parts of the book of direct interest. So, there is likely to be widespread appeal. In addition, academics with limited research experience, looking for a new research topic, will find here a rich mine of topics that need further study and research. This is so because, to deal with a question posed in litigation, I could proceed only thus far. What is described here is a good starting point, or points, for research. A prime example of this situation is Section 4.3 with the unusual title of 'The confused world of sulfate attack'.

About the arrangement of this book

This book is divided into seven chapters, each discussing a particular aspect of concrete. Chapter 2 deals with the water–cement ratio. Although this parameter has been recognized for a century, and even longer in France, there are still some aspects that are not clear or that are misused. I have tried to remedy that situation.

Chapter 3 deals with high-alumina cement. This is a specialty cement, valuable in some applications. It could, therefore, be thought that this cement need not be considered in a book on ordinary, run-of-the-mill concretes. However, much to my surprise, manufacturers of high-alumina cement periodically try to revive an interest in it as an ingredient of what would otherwise be ordinary concrete. I have been involved in experimental and technological aspects of high-alumina cement for fully half a century, and I am convinced that a revival of structural use of high-alumina cement is likely to lead to renewed failures and even to disaster. In my opinion, these attempts have to be nipped in the bud, and it is not enough to rely on the fact that the behaviour of high-alumina cement has been discussed in the past. As quoted in the opening paragraph of Section 3.3, 'Publications more than 25 years old are likely to be forgotten'.

Inadequate durability and shrinkage cracking are the major short-comings of concrete. Shrinkage is not discussed in this book because the subject has been thoroughly examined in the past. Although we can never claim that all that can be said on a particular topic has been said already, we do know enough about shrinkage to be able to control it in practice and to minimize its adverse consequences.

Durability covers a multitude of sins and, alas, we continue to sin. Sulfate attack, which is prevalent only in certain areas, is far from being understood. Section 4.3 tries to put the confused situation into some sort of perspective and identifies those aspects of sulfate attack that need clarifying. This would require a major and concerted effort, but no one is willing to face the challenge. Only fragmentary laboratory work is done, and often this is directed at confirming an individual researcher's hypothesis. What we need is a large-scale experimental investigation of *field* behaviour of concrete subjected to various sulfates in a wide range of situations. I very much hope that Sections 4.3 and 4.4 of this book will jolt someone into this truly worthwhile action.

The alkali–silica reaction has been studied for many years, yet Section 4.5 is devoted to that reaction. What provoked me into writing it is that I was recently called to investigate a potentially adverse situation. This led me to a study of recent European and British Standards prescribing design procedures to minimize the alkali–silica reaction in concrete.

Sections 4.1 and 4.2 have also arisen from a problem in practice, indeed in a court case. Progress of carbonation is the basic phenomenon considered and I use the hypothesis advanced in court to elucidate various aspects of carbonation in general.

Chapter 5 deals with two aspects of concrete in service. One is the intriguing question: which way do cracks run when concrete is subjected to stress? The second specific issue in Section 5.1 is whether coarse aggregate particles act as crack arresters and force the cracks which, following the straight line of maximum stress, would run through the particles, to deviate around them. While much theoretical work has been done, no experimental body of evidence is available.

The remaining two sections in Chapter 5, namely 5.2 and 5.3, deal with residential slabs on grade. These represent non-structural use of concrete because ground slabs in one- or two-storey homes are usually non-load bearing. Strength is thus of no import, and it might therefore be thought that there can be no great concern about the concrete mix used. And yet, in some parts of the USA, there has been a considerable amount of litigation about the alleged lack of durability of slabs.

To my knowledge, no substantial failures have been observed and no home has ever been condemned because of inadequacy of the slab on grade. Despite this situation, the allegations of non-conformity of the slabs with the American design codes have been advanced again and again, and out-of-court settlements have resulted in many millions of dollars being paid out. Such a situation is definitely harmful to the image of concrete but I could not address it while I was active as an expert witness. Now that I am saying a farewell to concrete, I am free to present my views in relation to three relevant codes of practice.

Section 5.2 deals with sustainability, but it does not pretend to be anywhere near an exhaustive treatment of this topic. Sustainability is concerned with economy of materials and energy, and also with longevity and durability of structures. Exhaustive treatment is still in search of an author. In the meantime, enthusiasts explore single aspects of durability. I shall do even less: I propose to consider concrete subjected to fire, and also to mention briefly the use of newer and new materials in the mix.

However, I shall also review the traditional attitude of designers with respect to economy, and thus to try to show that it has always been part of the designer's task to economize cost, materials, and energy. In other words, my approach to sustainability is more measured.

Chapter 6 has the umbrella title 'General issues'. Indeed, Section 6.1 deals with the entire topic of creating a concrete structure, starting with mix selection, all the way to the finished product. The importance of all the steps and procedures is emphasized. A complementary section, Section 6.2, looks at workmanship, a concept of considerable importance but not directly and objectively defined. Section 6.2 clearly distinguishes workmanship from design; this is obvious to engineers, designers and contractors but cement chemists occasionally muddy the waters, which is unhelpful.

A very short note on violation of codes is presented in Section 6.4. This is complemented by Section 6.3 re-visiting litigation in construction, this time in designed structures, as against non-designed slabs on grade, considered in Sections 5.2 and 5.3. Litigation is the bane of all those involved in construction. It won't go away, so that the best we can do now is to follow the old adage: forewarned is forearmed.

The last section in Chapter 6, Section 6.5, considers gender in the language dealing with concrete. It is like the proverbial red herring but we are enjoined to be politically correct at all times, even if the resulting syntax makes for difficult reading. Are we prepared to stand up and say that the pronoun 'it' covers a contractor, a designer, a tradesman and an operative regardless of sex, which is obviously irrelevant? I doubt that we are yet but in the fullness of time, let us revolt!

Chapter 7 presents an overview of the use and development of concrete over the last half century. Have we done well? Or well enough? Chapter 7 also contains a sombre farewell.

In Section 7.1, I look at the changes in concrete practice in the last 40 years, and I cannot make up my mind that it has all been progress, that is, a change for the better. Sections 7.2 and 7.3 look back on the 20th century. Section 7.4 attempts a glimpse of the future and includes suggestions for future developments.

And, finally, Section 7.5 gives my last word on concrete. The emphasis is on the personal pronoun 'my', which means that I have truly stopped writing – some might say pontificating – on concrete. Enough is enough.

2

Understanding the water–cement ratio

Water–cement ratio (w/c) is one of the oldest parameters used in concrete technology. Even those with a minimal knowledge of concrete are aware of the relation between w/c and the compressive strength and are, therefore, aware of the importance of w/c in that it is a primary influence upon strength. Strength or, strictly speaking, compressive strength at a specified age, usually 28 days, measured on standard test specimens, has traditionally been the criterion of acceptance of concrete.

The value of w/c is taken as a mass ratio. The mass of cement is an unequivocal term because cement is fed into the concrete mixer by weight. The mass of water has to be qualified by deciding *a priori* which kind of water we are concerned with. The kind of water can mean the water added into the concrete mixer which is available for the hydration of cement. This would exclude water absorbed by aggregate particles if these are less than saturated-and-surface-dry. But if there is free water on the surface of the aggregate, that is, water in excess of that required for the saturated-and-surface-dry condition, it is arguable that that excess water should be included in the mass of water taken into account in calculating the value of w/c.

Up to some 40 years ago, the above situation was taken into account by distinguishing *total w/c* and *effective w/c*. The latter value recognizes the saturated-and-surface-dry condition of the aggregate as the starting point for taking water into consideration. On the other hand, total w/c considers the whole water added into the mix. I believe total w/c was used in the old tests in the laboratory where the small quantities of aggregate could be brought to a desired condition, usually dry. In my opinion, the use of total w/c is unhelpful at best, or confusing, because there is no repeatable starting point for considering water as a component of the mix. Fortunately, the concept of total w/c fell out of favour; nowadays, we simply use the term w/c and presume that it considers water available for hydration.

The water available for hydration occupies space which, later on, will either be occupied by the products of hydration or will remain as a void. It is the void, that is the air in the concrete at the time of setting, that influences the strength of concrete. That much has been generally accepted, starting with Féret in France in 1892 and Duff Abrams in the USA in about 1919.

The above argument breaks down when w/c is smaller than would provide water necessary to achieve full hydration. This begs the question why the relation between w/c and strength applies at *all* values of w/c, including as low as 0.22, and why the relation is continuous. The problem is difficult and, seemingly, it had not been considered in the past. Section 2.1 offers some pointers and discusses also the influence of w/c on the rheological behaviour of concrete. We do not claim, however, that Section 2.1 represents the last word on the subject; that section is an excellent starting point for a highly competent researcher.

Section 2.2, by contrast, deals with hardened concrete. The value of w/c is established by the input into the mixer. Once mixing has taken place, how do we determine the value of w/c? In practice, the question arises when the concrete has hardened, and usually is many months or years old. Why is this so?

The maximum value of w/c is often prescribed in a specification because this is relevant to durability. Although, as discussed in Section 2.1, the value of w/c is closely related to the compressive strength of concrete, it is the strength as such that is explicitly specified. Often the value of w/c necessary to achieve the specified strength is higher than the value prescribed from durability considerations. If the producer chooses the mix proportion so as to satisfy strength and ignores the w/c for durability, then how do we verify the situation in concrete in situ?

The short answer is that no test for a direct determination of w/c of hardened concrete is available. An indirect approach is possible by determining the cement content in the concrete (in kg/m^3) and also estimating the original water content (in litres per m^3) from the volume of voids in the hardened concrete and the volume of hydrated cement paste (knowing the volume of the water of hydration). The errors associated with each of these measurements, especially of the cement content, unless a sample of the original cement used is available, are large. Cumulatively, the errors mean that the calculated w/c is within ± 0.05 of the specified value. In other words, if the maximum w/c was specified to be, say, 0.45, and the tests on hardened concrete lead to an estimate of 0.45, the 'true' value is somewhere between 0.40 and 0.50. Is the concrete, therefore, compliant or non-compliant?

The issue may become the subject of a court case, and has indeed done so on many occasions. I have appeared as an expert witness in several cases, and this has forced me to review the entire issue of the accuracy and precision of the determination of the value of w/c on a sample of hardened concrete. The upshot is that a maverick claim of so-called green-tone fluorescence has not been upheld. This test relies on epoxy impregnating a very small sample of concrete with a low-viscosity resin containing a fluorescent dye, which fills the capillary pores and other voids in the concrete. A thin section of impregnated concrete is viewed in a transmitted ultraviolet light, and its colour is compared by eye with a sample of concrete with a known w/c.

So far so good, but if no reference sample is available (because, at the time of placing the concrete, no preparations for a court case many years later would have been made) then what? One chemist has claimed that he can 'tell the colour' with great accuracy, and hence pronounce on the original w/c.

While there are piano tuners with an infallible ear, and Leonardo da Vinci may have been able to tell a hue of green (and this is the colour in the test) in steps of w/c of 0.01, this is hardly the basis of a scientific test, the consequences of which – compliance or non-compliance with the specification – may result in large sums of money being paid or received.

It is of interest that in several cases the Superior Court of California ruled that the test is not acceptable as evidence; nevertheless the proponent of the test continues to advance its use. I should add that these Court decisions are not binding in other cases, but they show the judge's thinking. This is why one such decision is given in Section 2.3.

In practical terms, the issue is rarely that of compliance or non-compliance of a tiny sample of concrete. It may appear to some chemists that concrete in situ is a uniform material so that testing a tiny sample established the properties of 'the concrete'. Engineers, however, are aware of the heterogeneity of concrete in situ. Placing and compacting induce variability in addition to that due to the 'natural' variability of the mix.

Thus, while compliance at the point of discharge from the mixer may be a legitimate issue between the supplier and the user of concrete, consideration of compliance after transportation and after curing, or its absence, and after a period in service is highly complex.

In summary, Section 2.2 presents a careful review of the determination of the w/c of hardened concrete, and confirms that a useful determination is not practicable.

2.1 HOW THE WATER–CEMENT RATIO AFFECTS CONCRETE STRENGTH

P.-C. Aïtcin and Adam Neville

This is not just another paper on the water–cement ratio (w/c): it goes beyond the well-known relation between w/c and strength and tries to get behind that relation. In other words, our intention is to try to find out whether the influence of the w/c on strength arises simply from the fact that a part of the mixing water produces voids that weaken the mass of hydrated cement paste or whether there are other, possibly more fundamental, factors involved. To resolve the problem, we have to consider not only the hydrated cement paste but also mortar or concrete containing this cement paste. In this connection, it is useful to look into the relation between the strength of mortar and of concrete, both containing the same cement paste. To start with what is well-known, we shall first consider the early studies on the w/c.

Relation between strength and w/c

For the sake of brevity, the term 'strength' will be used to mean compressive strength. Probably the first formulation of the relation between strength and the nonsolid ingredients of concrete was made by Féret in France in 1892 [1]. He understood the fact that the presence of water- and air-filled space in mortar has a negative influence on strength and, from his experimental work, established a power relation

$$f'_c = kC^2/(C + W + A)^2 \qquad (1)$$

where C, W and A each represent the volume of cement, water and entrapped air respectively, in a unit volume of concrete; and f'_c is the compressive strength of mortar determined on test specimens of specific shape and size at a specific age. Féret found that a power of 2 gave the best fit and determined the value of the coefficient k by experiment.

Still working in terms of volumetric proportions, by introducing the term W/C, we can rewrite Eq. (1) as

$$f'_c = k/[1 + W/C + A/C]^2 \qquad (2)$$

It may be thought unusual to deal in terms of volumetric proportions, but this is the only scientifically sound approach. The volumetric

11

proportions of the relevant ingredients in 1 m^3 of concrete are typically: $C = 0.1$, $A = 0.01$ to 0.02, and $W = 0.2$. Thus, $W/C = 2$ and $A/C = 0.1$. We can, therefore, neglect the term A/C and rewrite Eq. (2) as

$$f'_c = k/[1 + W/C]^2 \tag{3}$$

Given that in reality concrete is batched not by volume but by mass (possibly with the exception of water), and that engineers usually work in terms of mass proportions, with c and w representing the mass of cement and water, respectively, in a unit volume of concrete, we can write Eq. (3) as

$$f'_c = k/[1 + 3.15w/c]^2 \tag{4}$$

This equation recognizes the fact that the specific gravity of water is 1, the typical specific gravity of Portland cement is 3.15, and uses the universal term for the water–cement ratio by mass, w/c.

A relation between strength and the W/C was developed, probably independently, by Abrams [2] in the USA as

$$f'_c = A/B^{W/C} \tag{5}$$

where W/C is expressed by volume, and A and B are constants dependent on the specific conditions such as the cement used, curing, and age at test.

The work of Féret and of Abrams represents a significant contribution to the understanding and, above all, use of concrete, and for that they deserve considerable credit. The cements they used, however, were very different from modern cements and, even more importantly, modern concretes usually contain additional ingredients. Their cements were much more coarsely ground than modern cements, and their chemical compositions were different. Moreover, Féret and Abrams used no plasticizing admixtures – indeed no admixtures at all. Thus, the initial reactivity of their cements was low, so that the workability of concrete was almost solely governed by the water content in the mix.

As always, the concrete used by Féret and Abrams had to have an adequate workability to achieve reasonably full compaction. The *absolute* need for adequate workability must never be forgotten by those who proportion concrete mixes in the laboratory. Because cement was an expensive ingredient, the cement content in the mixes used was relatively low, so that a high water content entailed a

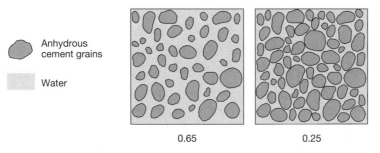

Fig. 2.1.1. Diagrammatic representation of fresh cement paste with a w/c of 0.65 and 0.25. The ratio of the areas of water and cement is equal to the W/C by volume

high w/c. Typical values used by Féret and Abrams were between 0.6 and 1.0 by mass. A visual appreciation of the difference in the spacing of cement particles at a w/c of 0.65 and of 0.25 can be obtained from Fig. 2.1.1.

Water requirement for hydration

Early in the 20th century, the knowledge of cement hydration was largely that established by Le Chatelier, and it was only in the middle of that century that Powers and his co-workers established quantitative data about water involved in the hydration of cement and about the volumes of cement, water, and products of hydration. The findings of Powers are valid to this day but, surprisingly, some engineers are still not entirely clear about the water used up in the hydration of cement and the water necessary for the hydration to proceed.

The source of confusion lies in the fact that the water required to fully hydrate 1 gram of C_3S or C_2S is about 0.22 gram. The volume of products of hydration of Portland cement, however, is larger than the sum of the volumes of the cement and water participating in the reaction. Specifically, hydrated cement paste contains about 30% of very fine pores, known as gel pores. The gel pores must remain filled with water.

It follows that a mix with a w/c of 0.22 cannot hydrate fully. Indeed, a w/c of not less than about 0.42 is necessary for full hydration to be possible. This arises from the fact that it requires a volume of 1.2 mL to accommodate the products of hydration of 1 mL of anhydrous cement. In other words, the minimum W/C by volume for complete hydration is 1.2. This is equivalent to a w/c of about 0.42 by mass.

13

To put it another way, the minimum mass of water necessary for full hydration is twice the mass required stoichiometrically for the formation of calcium silicate hydrates. A further complication arises from the fact that the water in the gel pores, which is adsorbed on the surface of the gel, has a higher density than free water, namely, about 1.1.

Some engineers are bothered that using a w/c lower than 0.42 means that some cement will always remain unhydrated; others are unhappy that at a w/c of 0.42, or higher, some capillary space is always present in the hydrated cement paste.

These issues are discussed in *Properties of Concrete* [3] but seemingly the need for better understanding continues. As recently as May 2002, an article on this topic was published by Mather and Hime [4], but a fuller discussion here may be beneficial. First of all, the figures given in this article are only approximate, and they do not recognize the fact that the different compounds in clinker result in different products of hydration, with varying amounts of chemically bound water. The size of cement grains, that is, the fineness of cement, also influences the rate of hydration, as well as the degree of hydration, achieved in practice. It is worth repeating that the cements used by Féret and Abrams were very much coarser than modern cements.

As for some part of the original cement grains remaining unhydrated, it should be pointed out that these remnants are by no means harmful. Indeed, they can be considered as very fine aggregate, admittedly much more expensive than conventional aggregate, but with the benefit of excellent bond to the products of hydration.

Water requirement of concrete
Considering further the concretes in the era of Féret and Abrams, we should note that in the absence of any plasticising admixtures, they required a w/c much higher than 0.42 or W/C much higher than 1.3. We can, therefore, subdivide in Eq. (3) the term W/C into two parts: W_h/C, where W_h represents the water required for full hydration to be possible; and W_p/C, where W_p represents the *additional* water for an adequate workability, which could be called 'workability water'. We can, therefore, write Eq. (3) as

$$f'_c = k/[1 + (W_h + W_p)/C]^2 \tag{6}$$

Considering $W_h/C = 0.42 \times 3.15 = 1.32$, we obtain

$$f'_c = k/[2.32 + W_p/C]^2 \tag{7}$$

14

which, in terms of mass proportions, and with a different coefficient k', is

$$f_c' = k'/[0.74 + w_p/c]^2 \tag{8}$$

By separating out the water that takes no part in the formation of the hydrated cement paste, either by being chemically combined or adsorbed, Eq. (7) links the strength of concrete to the porosity of the gross volume of hydrated cement paste. Thus, Eq. (7) represents the relation between the compressive strength of concrete and the porosity of the hydrated cement paste. This lies at the root of the 'water–cement ratio rule' that engineers have been using for three-quarters of a century. Explicit reference to porosity as a factor governing strength was made by Sandstedt, Ledbetter and Gallaway in 1973 [5].

The influence of porosity of hydrated cement paste on strength accords with the known increase in strength as hydration progresses and, because continuing hydration results in a greater volume of solids than that originally occupied by the hitherto unhydrated part of cement, the progress of hydration results in a decreased porosity.

A further observation on the influence of porosity on strength is provided by the effect of air entrainment on strength. It is well

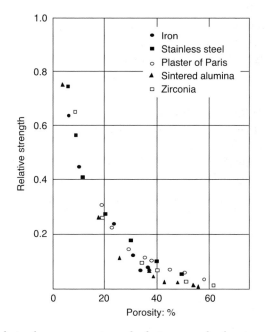

Fig. 2.1.2. Relation between porosity and relative strength of various materials [3]

15

known that, at a constant w/c, every 1% of air in concrete reduces its strength by about 5%.

Such a relation is not unique to concrete. This was pointed out in 1981 by Neville in the 3rd Edition of *Properties of Concrete*, where it is shown that the same curve represented the relationship between the relative strength and porosity for materials as diverse as concrete, iron, stainless steel, plaster of Paris, sintered alumina, and zirconia; the 4th Edition substantially repeats the material [3] (Fig. 2.1.2). Aïtcin also developed these concepts in his book published in 1998 [6].

Mixes containing silica fume

As far as hydrated cement paste is concerned, we have considered Portland cement only. Rao [7] conducted experiments on mortars with different contents of silica fume and reported that their compressive strengths at different ages did not follow the classical expression relating strength to w/c. We find this observation not surprising because the chemical reactions of silica fume do not progress at the same rate as the reactions of Portland cement. In consequence, at very early ages, silica fume can be considered as an inert, very fine material and not as a binder.

The situation is complicated, however, because of the physical effects of silica fume in that it affects the packing of particles and, therefore, the forces controlling the proximity of cement particles to one another at the beginning of the hydration process. Moreover, the very fine particles of silica fume act as nucleation sites for the hydration of cement.

Furthermore, the chemical role of silica fume in the process of hydration is governed by the availability of $Ca(OH)_2$, derived from the hydration of C_3S and C_2S in Portland cement. It follows that, above a certain threshold in the binder, silica fume cannot undergo a significant amount of chemical reaction, so that the 'excess' silica fume is more truly an inert filler rather than a binder.

Factors other than w/c

The discussion so far, which translates W/C into porosity, justifies the 'water–cement ratio rule', but it does not entirely explain the underlying physical situation. In any case, it should be pointed out that it is only pores above a certain minimum size that influence strength, so that the pore size distribution also enters the picture. It is intuitively understandable that more voids mean a reduction in strength, but strength must derive from bonds, chemical or physical, in the mass of concrete. How do we reconcile the 'water–cement ratio rule' with the role of the bonds?

A second question concerns the validity of the water–cement ratio rule for values of w/c between 0.42 and 0.22, given that, for these values of w/c, all the originally water-filled space will have been filled by the products of hydration of cement when maximum hydration has occurred.

A third question concerns any possible difference between the strength of hydrated neat cement paste and the strength of mortar and concrete containing the same paste.

To try to answer the first question, we should return to the concept of workability water. The need for this water in Portland cement mixes without any admixtures arises from the presence of electrical charges upon the surface of anhydrous cement particles. These charges cause flocculation of cement particles, which has a negative effect on workability and impedes early hydration.

Unlike the situation half a century ago, it is now possible to defloccu-late cement particles by means of inclusion of superplasticizers (high-range water-reducing admixtures) in the mix. These specific organic molecules are very effective in neutralizing the electrical charges on the surface of cement particles, either by an electrostatic process (as in the case of polynaphthalene, polymelamine, and lignosulfonates) or by steric hindrance (as in the case of polycarboxylates and poly-acrylates). The effect of superplasticizers is to liberate water that would otherwise be trapped in the flocs of cement particles. In conse-quence, a smaller amount of mixing water will provide the required workability.

It is thus possible to make workable concrete with a w/c down to a value as low as 0.22, that is, without any workability water. Hence, Eq. (4) would not be expected to be valid – and yet it is valid!

Influence of transition zone

A study of concretes made with high values of w/c, such as those used by Féret *et al.*, reveals that the porosity of the hydrated cement paste in the vicinity of aggregate, mainly the coarse aggregate, is higher than the porosity of the paste further away. The zone near the interface paste-aggregate is called the transition zone or interface zone. The hydrated cement there has been formed by the process of solution of the cement compounds and subsequent precipitation; such hydrated cement is sometimes referred to as external hydration products.

On the other hand, away from the interface zone, hydration takes place largely by diffusion of water molecules into the unhydrated cement; the resulting hydrates are sometimes referred to as internal hydration products.

We should note, in passing, that the transition zone does not exist in hydrated neat cement paste and is present to a much smaller extent in mortar. The differences between the strengths of mortar and of concrete will be discussed in a later section.

The extent of the transition zone has been found to depend on the w/c of the mix; at very low values of w/c, the transition zone is absent because there is little water-filled space to permit solution and precipitation. In consequence, hydration proceeds by diffusion only. An interpretation of this situation is that, at very low values of w/c, the influence of the w/c on strength is not due to the presence of capillary pores created by hydration but through the quality of the hydrated cement paste. The two influences on the w/c are, however, quite closely connected, as will be shown in the next section.

Influence of bonds

The preceding discussion permits us to address the issue of bonds within the hydrated cement paste in relation to compressive strength. We have just shown that bonds, as well as porosity, influence strength. That the development of bonds is more difficult in mixes with a high w/c can be seen from Fig. 2.1.1. This figure shows that, at a w/c of 0.65, the products of hydration have to extend a considerable distance to create inter-particle bonds that give the hydrated cement (and concrete) its strength. Thus, high w/c directly influences the development of bonds, both in slowing down the increase in strength and in its ultimate value.

Although we have mentioned the influence of porosity on strength separately from the influence of bonds, they are not distinct; a higher porosity means weaker bonds because of the distances over which they need to develop. Thus, bonds are the fundamental factor affecting strength.

We believe that the concept of bonds as a fundamental factor affecting strength, with the w/c influencing the strength of bonds, is an important contribution to the understanding of the strength of concrete at a practical level. This concept puts the w/c rule in its proper place.

Nevertheless, we have to admit that we are avoiding a closer definition of the term 'bonds'. They could be adhesion and cohesion forces. Or the quality of the bonds could depend on the absence of micro-cracking developing at the time of early hardening, following setting. Or their quality could be the absence of differential deformation due to shrinkage. Alternatively, the quality of bonds could be adversely affected by differences in the modulus of elasticity, Poisson's ratio, and the coefficient of thermal expansion between the hydrated cement paste and the aggregate. These questions remain to be answered, but they are more in the domain of science than engineering.

Rheological aspects of w/c

We have seen how the value of w/c influences the development of bonds. Their development in the period immediately following mixing has a direct effect on the workability of the mix. The traditional approach to ensuring an adequate workability is to provide some workability water, that is, to have large spaces between the particles of unhydrated cement.

At the same time, to ensure reasonably rapid construction requires the development of some early strength. This can be satisfied by the use of cement with a chemical composition such that some crystals grow fast; these are the products of hydration of C_3A, namely ettringite. They do not reduce workability because the cement particles are far apart. The crystal growth in cements having a high C_3A content in mixes with a high w/c is shown diagrammatically in Fig. 2.1.3.

Now, the use of cements with a high C_3A content in mixes with a low w/c is unsatisfactory: it can be imagined that a rapid growth of hydrates between closely spaced particles of cement leads to a rapid loss of workability. Using scanning electron microscopy, however, we have not observed needles of ettringite. Also, crystals of C-S-H are sparse, the appearance of C-S-H being rather that of an amorphous mass. We are aware, however, that other investigators did see ettringite in a similar situation.

In the latter part of the 20th century, and in response to the demand for early development of strength, the cement manufactured had a high content of C_3S, as well as of C_3A, and also a high fineness. Such cements have serious disadvantages, in addition to too rapid a loss of workability. Briefly, these disadvantages are: development of high, early heat of hydration, increased shrinkage, increased creep (which may or may not be harmful), a higher risk of sulfate attack, risk of

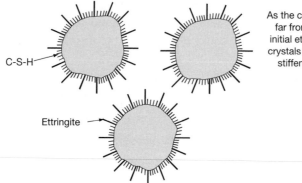

C-S-H

Ettringite

As the cement particles are far from each other, the initial ettringite and C-S-H crystals cause only a slight stiffening of the paste

(a) Early hydration – dormant period

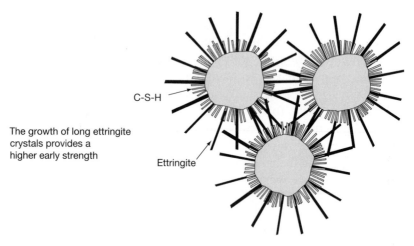

C-S-H

The growth of long ettringite crystals provides a higher early strength

Ettringite

(b) Setting and hardening (crystals of Ca(OH)$_2$ not shown)

Fig. 2.1.3. Schematic representation of the hydration of a paste with high w/c made with Portland cement with a high content of C_3A

delayed ettringite formation (if concrete is subjected to steam curing) and possibly higher carbonation.

At low values of w/c, a high content of C_3A is unnecessary because early development of strength is ensured by products of hydration of C_3S and C_2S, namely C-S-H, overlapping between adjacent cement particles. Achieving a satisfactory workability at low values of w/c, however, necessitates the use of high-range water-reducers in the mix. The use of cement with a low content of C_3A results in concrete

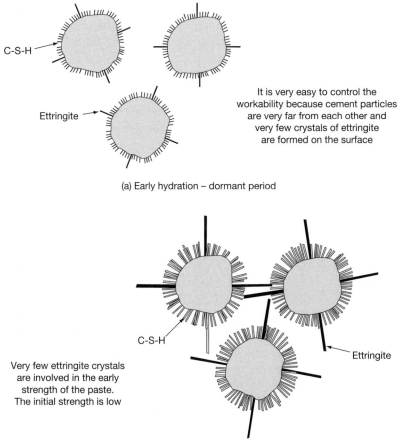

It is very easy to control the workability because cement particles are very far from each other and very few crystals of ettringite are formed on the surface

(a) Early hydration – dormant period

Very few ettringite crystals are involved in the early strength of the paste. The initial strength is low

(b) Setting and hardening (crystals of Ca(OH)₂ not shown)

Fig. 2.1.4. Disadvantages of Portland cement with a low C₃A content in mixes with a high w/c

having a low porosity and a lower risk of sulfate attack. It should be emphasized that cements with a low content of C_3A are unsatisfactory at high values of w/c if high early-strength development is required (Fig. 2.1.4).

In other words, cements with high contents of C_3A and C_3S are desirable for mixes with a high w/c, but the use of such cements may entail disadvantages with respect to durability. On the other hand, for mixes with a low w/c, it is preferable to use cements having low contents of C_3A and C_3S but as already mentioned, a superplasticizer must be incorporated in the mix.

Strength of mortar and of concrete

This section, so far, has dealt principally with the strength of hydrated cement paste, but we mentioned the influence of the transition zone on the properties of hydrated cement in concretes with a w/c higher than about 0.40. This brings us to the question posed earlier: Is there any effect of aggregate on strength? In other words, does the relation between strength and w/c considered so far for hydrated cement paste apply equally in the presence of aggregate?

Concretes are made with a wide variety of aggregates and, generally, the strength of the aggregate is not considered. The assumption is that the compressive strength of aggregate *per se* does not affect the strength of concrete, provided that the strength of aggregate is higher than the strength of hydrated cement paste, and is also higher than the bond between the hydrated paste and the aggregate particles. Admittedly, the latter is not measured directly, but it is evidenced by the mode of failure of concrete specimens in a compression test: if the failure takes place largely at the aggregate-paste interface, the bond is taken to be weaker than the strength of the hydrated cement paste, and also weaker than the strength of the aggregate. Thus, looking at a specimen after it has been tested in compression is instructive. It is only fair to add that this is a qualitative assessment because the actual stresses involved are not properly considered (see Section 5.1).

The preceding discussion might be thought to lead to the conclusion that the strength of concrete is governed solely by the strength of the hydrated cement paste. A corollary of such a situation would be that the strength of the hydrated cement paste with a given w/c is the same as the strength of concrete containing that hydrated cement paste.

Whether or not this is so has not been reliably established. One reason is that it is very difficult to make test specimens of neat cement paste, except very small ones, because of the rise in temperature upon hydration of cement. On the other hand, concrete specimens require the least dimension that is a multiple of the maximum aggregate size; for example, ASTM C 31 prescribes a ratio of 3.

More generally, it has been accepted for quite some time that, in order to characterize the strength of cement, tests are performed on specimens of mortar or of concrete containing that cement and not a neat cement paste. The interaction between cement and concrete in practice has been studied by Neville [8].

Let us now look at the strength of mortar as compared with the strength of concrete, both materials having the same w/c. It is interesting to note the comment by Struble [9] to the effect that evidence for relying on the strength of cement paste at the same w/c and degree of hydration to predict the strength of concrete is 'generally lacking'. Also, as reported by Gaynor [10], ASTM (American Society for Testing and Materials) Committee C 01 expressed the view: 'Caution must be exercised in using the results of this method (ASTM C 109 to measure the strength of mortar) to predict the strength of concretes.' Gaynor further says: '...cement manufacturers are reluctant to assume that these strengths [of mortar] bear any formal relationship to strength performance in concrete'. Nevertheless, Gaynor concludes his paper by saying that there is a correlation between tests on mortar and the performance of cement in concrete. The term 'performance' is not quite the same as strength.

On the other hand, Weaver *et al.* [11], who are cement manufacturers, say about their study: 'No correlations of strength from any of the mortar tests with the 28-day strength of concrete were found in the study, even when the concrete was made to the standard mix proportions'. An explanation of this statement was offered by Neville [8]. With admixtures and different cementitious materials, the relation is likely to be more problematic.

There is also a difference between mortar and concrete in that, in the latter, the relative volume of hydrated cement paste is lower than in the former. This means that the relative volume of voids in concrete, that is, its porosity, is lower than in the mortar. Conceptually, we could expect, therefore, the strength of concrete to be higher. Thus, the aggregate content in the mix should affect it strength. But is this so? No conclusive evidence is available, and what there is, is unclear – not to say confusing.

We are not looking for *equality* of strength of mortar and concrete, if only because the test specimens used for the two materials are not of the same shape and size, and it is known that shape and size affect the strength determined by test. The characteristics of test specimens do not affect the inherent strength, but *inherent* strength is an elusive concept.

It is useful to add that the characteristics of the test specimens influence the temperature rise of mortar or concrete caused by the exothermic reactions of hydration of cement. To remove the temperature effect, tests can be conducted under isothermal conditions. Such tests were performed by Carino and Tank [12]. In their experiments, they determined the rate constant for strength development at the reference temperature of 23 °C, with k_r expressed in $(\text{day})^{-1}$. This parameter is used in their equation for relative strength, that is, the ratio of strength S at the equivalent age at reference temperature (t_e) to the compressive strength at infinite age, S_u. Using the symbol t_{or} for the age at the start of strength development at the reference temperature, they express relative strength as

$$S/S_u = k_r(t_e - t_{or})/[1 + k_r(t_e - t_{or})] \tag{9}$$

Carino and Tank tested seven mortar and concrete mixes, with values of w/c of 0.45 and 0.60, made with Portland cement Types I, II and III alone, and with Type I cement together with one of the following: fly ash, ground granulated blast-furnace slag, an accelerator and a retarder. What is relevant to the present study is that, considering all these tests at both values of w/c, they found a good linear relation between the values of the rate constant for strength development, k_r for concrete and for mortar, but the values of k_r for concrete are about 18% greater than those for mortar. This means that, for equal long-term strength, the initial rate of strength development in concrete is higher than in mortar.

The objective of the work by Carino and Tank [12] was to make it possible to estimate the relative strength development of concrete cured at different temperatures from tests on mortar with the same w/c. The conclusion that we can draw from their work is that the relative strength development is independent of the w/c (0.45 or 0.60). This finding suggests that the presence of coarse aggregate does not affect the *rate* of strength development. However, there may be an influence of the coarse aggregate on the *value* of strength, but this cannot be established from the tests by Carino and Tank.

In passing, we can mention that very limited tests by Kolias [13] have shown correlations between the porosity of the cement paste and the strength of mortar, and between the porosity of the cement paste and the strength of concrete; but the two correlations are not the same.

Conclusions

We have reviewed the well-known relationship between compressive strength of concrete and the w/c, which is based on porosity, and we have also recognized that porosity is not a factor in this relationship at very low values of w/c. It follows that bonds within the hydrated cement paste must be the governing influence on strength. The considerations of porosity and of bonds can be combined by appreciating that porosity affects the strength of bonds through the influence of the distance between hydrating surfaces. Thus, a single relation between strength and w/c is based on the concepts of bonds as a fundamental factor.

Our study of the relation between the strength of mortar and the strength of concrete, both materials containing a cement paste with the same w/c, has failed to establish a physical explanation of the available test results.

The w/c influences the rheological behaviour of concrete. At high values of w/c, that is, when no superplasticizers are included in the mix, a higher workability and a reasonably rapid development of strength are obtained using cements with high contents of C_3A and C_3S; however, such cements may lead to undesirable properties of concrete in terms of durability. On the other hand, at low values of w/c, when superplasticizers are used, cements with low contents of C_3A and C_3S lead to the production of concrete with a satisfactory workability, early strength and durability; we are still studying the mechanisms involved.

Finally, understanding how the w/c affects the strength of concrete is not just a matter of scientific curiosity. Such understanding should help us to choose suitable mix proportions and thus achieve better concrete in practice.

References

1. Féret, R., Sur la compacité des mortiers hydrauliques, *Annales des Ponts et Chaussées*, 2nd semester, 1892, pp. 5–61.
2. Abrams, D.A., Design of concrete mixtures (1925), *A Selection of Historic American Papers on Concrete 1876–1926*, SP-52, H. Newlon, Jr, Editor, American Concrete Institute, Farmington Hills, MI, 1976, pp. 309–330.
3. Neville, A.M., *Properties of Concrete*, Fourth Edition, Longman & John Wiley, London and New York, 1995.
4. Mather, B. and Hime, W.G., Amount of water required for complete hydration of Portland cement, *Concrete International*, 24, 6, June, 2002, pp. 56–58.
5. Sandstedt, C.E., Ledbetter, W.B. and Gallaway, B.M., Prediction of concrete strength from the calculated porosity of the hardened cement paste, *ACI Journal Proceedings*, 70, 2, Feb., 1973, pp. 115–116.
6. Aïtcin, P.-C., *High Performance Concrete*, E & FN Spon, 1998.
7. Rao, G.A., Role of water–binder ratio on the strength development in mortars incorporated with silica fume, *Cement and Concrete Research*, 31, 2001, pp. 443–447.
8. Neville, A., *Neville on Concrete*, Section 4.1, American Concrete Institute, Farmington Hills, MI, 2003.

9. Struble, L., The performance of Portland cement, *ASTM Standardization News*, Jan., 1992, pp. 38–45.

10. Gaynor, R.D., Cement strength and concrete strength – an apparition or a dichotomy?, *Cement, Concrete, and Aggregates*, ASTM, 1993, pp. 135–144.

11. Weaver, W.S., Isabelle, H.L. and Williamson, F., A study of cement and concrete correlation, *Journal of Testing and Evaluation*, 2, 4, 1974, pp. 260–303.

12. Carino, N. and Tank, R.C., Maturity functions of concrete made with various cements and admixtures, *ACI Materials Journal*, 89, 2, Mar.–Apr., 1992, pp. 188–196.

13. Kolias, S., Investigation of the possibility of estimating concrete strength by porosity measurements, *Materials and Structures*, 2, 1994, pp. 265–272.

DISCUSSION

Letter to the editor: Water–cement ratios for lean RCC

Regarding the article 'How the water–cement ratio affects concrete strength', by Pierre-Claude Aïtcin and Adam Neville (CI, August 2003), this most interesting account about the relationship between strength and water–cement ratio (*w/c*) relates to the rheological behaviour of conventional concrete. The inverse relation between strength and *w/c does* hardly apply for lean roller-compacted concrete (RCC) mixtures. I have two questions for the authors about the relevance of their finding on lean RCC.

1. Water–cement ratios for lean RCC are between 2.0 and 1.0. These numerically high *w/c* ensue from the low value of cementitious material (and not from the high water content). Decreasing *w/c* will reduce the paste volume necessary for compaction. If this is done to an extent where air voids are not filled, strength will be reduced. In other words, for lean RCC mixtures there is the need for excess water for hydration because the water required for hydration is low and the excess water is needed to provide moisture for compaction. How does this 'mundane' requirement affect the bond, as discussed by the authors?

2. The constituents of the lean RCC paste (±20% by volume) are different than in conventional concrete: the fines lacking from the low content of cementitious materials have to be replaced by fillers (aggregate fines) that do not participate in hydration. Does this situation affect the authors' 'transition zone'?

Harald Kreuzer
Brunegg, Switzerland

Authors' response

Mr Kreuzer's questions present an interesting extension of our article. They seem to arise from the fact that, in our article, we did not explicitly state that our arguments apply only to conventional concrete, that is, dense concrete in which the entrapped air content is minimal.

We view the situation as follows. Lean roller compacted concrete (RCC) is more related to compacted soil than to usual concrete. In lean RCC mixes, the volume of the paste is not sufficient to fill all the voids that exist between the aggregate particles after their compaction. In such mixes, the amount of air voids cannot be neglected, not only from the volumetric point of view, but also from micro-mechanical considerations, as it can be in conventional concrete.

In lean RCC, due to the lack of paste, not all the surface of the aggregate particles is surrounded by the paste and there are numerous aggregate-to-aggregate contact areas. Thus, it is difficult to define a typical transition zone, as can be done in a conventional concrete. After hardening of the paste, inter-aggregate arching is developed by the paste with significant consequences. For example, in lean RCC, the ratio of the modulus of rupture to compressive strength is about twice as high as in conventional concrete.

Finally, the w/c ratio of lean RCC mixes is usually calculated on the basis of dry aggregate, and not on the basis of saturated aggregate, as in conventional concrete. Due to the high content of aggregate in lean RCC mixes, this makes a significant difference.

The point brought out by Mr Kreuzer shows that the so-called w/c is not a universal 'law' applying to any type of material containing some Portland cement, but rather a relationship that works well in conventional concrete where the hydrated cement paste constitutes a continuous medium between the aggregate particles.

We are grateful to Mr Kreuzer for his questions, which gave us an opportunity to make the above remarks.

Pierre-Claude Aïtcin
Adam Neville

2.2 HOW CLOSELY CAN WE DETERMINE THE WATER–CEMENT RATIO OF HARDENED CONCRETE?*

Abstract

Generally, a determination of the water–cement ratio (w/c) of hardened concrete is not very useful, but there exist indirect methods for that purpose. A direct method, utilizing optical fluorescence microscopy, has been standardized by a Nordic organization; it requires the use of reference standards of a range of mixes with specific ingredients and a known curing history. The method is useful for quality control in repetitive production of concrete elements.

A recent paper claims that the w/c of unknown concrete can be determined with a precision of ± 0.02. A review of that paper, utilizing simple statistical tests, shows that the claim is unfounded, so that caution in the use of optical fluorescence microscopy is required.

Introduction

Water–cement ratio (w/c) is generally used as an important parameter in the design of concrete mixes, mainly because of its strong influence on the strength of concrete but, nowadays, its significance is not as great as it was in the past [1].

Consideration of the w/c leads generally to prescribing its value in the specification and subsequently to instructions about the quantities of water and of cement in batching. Once the concrete has been produced, its properties that are commonly measured are workability (usually established as slump) and air content (if air entrainment is used), and thereafter tests are performed on hardened concrete. The most common of these is compressive strength and it is accepted that, if there is no unexpected reduction in strength, the w/c continues to be as specified.

The determination of the w/c of hardened concrete is highly uncommon. It is, therefore, not surprising that there exists no test for w/c either in international standards (ISO, EN or RILEM) or in American standards (ASTM). Nevertheless, there arise situations where it is desirable to establish the value of w/c of hardened concrete, for example when non-compliance is suspected.

---

* Originally published in *Materials and Structures*, 36, June, 2003 and reproduced here with kind permission of Springer Science and Business Media.

Possible procedures using existing indirect methods are considered in Part I of this section. Part II deals with the methods of optical fluorescence microscopy. Part III discusses a specific claim of achieving a high precision by the use of the optical fluorescence microscopy technique.

Part I: Existing methods

As I have already stated, there exists no standardized and recognized method of determining the w/c of a sample of concrete taken from an existing structure. The best that can be done is to estimate separately the numerator and the denominator in the term w/c.

Chemical methods

The value of the denominator c is established as the cement content by mass in a sample of concrete of unit volume, that is as kilogrammes of cement in 1 cubic metre of concrete. No accepted direct method is available, and recourse is usually made to chemical methods for the determination of soluble silica and calcium oxide. Details of the calculations are outside the scope of the present book, but what is relevant is that the precision is low, especially in mixes with low cement contents. For example, British Standard BS 1881:Part 124:1988 (which is still in force) gives the value of reproducibility as between 50 and 60 kg/m^3. This means that there is a 95% probability that the difference between two single test results found on identical samples of the same concrete by two analysts working in different laboratories at different times will not be greater than 50 to 60 kg/m^3. Viewed against cement contents of 250 to 350 kg/m^3, these are large 'errors'. Now, the water content, that is the numerator in the term w/c, is estimated as the sum of the mass of chemically combined water in the hardened cement paste and of the volume of capillary pores, which represent the remainder of the original water in the mix. Details of the determination of the water content in a sample of hardened concrete are outside the scope of the present section, but some caveats in British Standard BS 1881:Part 124:1988 should be noted. These include the following injunctions: the concrete must be sound and not damaged either physically or chemically; it must not be poorly compacted or air-entrained; and it should not be older than five years.

From the two separate determinations of w and c, the value of w/c is calculated by a simple division. It is not surprising that the precision of the outcome is low: according to the *Concrete Society Technical Report*, the calculated value is likely to be within 0.1 of the actual w/c in the mix [2]. Even this precision is thought by St John *et al.* to be optimistic; they say: 'in practice, the determined values of original water–cement ratio by this method should be treated with considerable circumspection, even when obtained by experienced analysts' [3].

Physical methods

For the sake of completeness, I should refer to the use of the linear-traverse method and the modified point-count method, which allow a determination of the volumetric composition of a sample of concrete by optical microscopy techniques; these are given in ASTM C 457. The volumes of aggregate and of voids (containing air or evaporable water) are determined, and the remaining volume is assumed to be hydrated cement. Given that hydrated Portland cement paste contains a fixed proportion of non-evaporable water, the original content of cement in the mix can be calculated. The cement content is determined within 10%. However, because the test cannot distinguish between air voids and voids that were originally occupied by water, the original water can be estimated only very approximately. Hence, any calculation of the original w/c is of minimal value.

It is even more difficult to determine the w/c of concrete made with high-alumina cement, and Bate [4] said that in one particular structure the free w/c was estimated to be 'in the range 0.45–0.55', that is, within ± 0.05. He added: 'various methods of estimating free w/c were tried. It must be emphasized, however, that none gave a precise determination and all were indicative only' [4].

Attempts have been made to determine the w/c of hardened Portland cement paste (not concrete) by impedance spectra, but these are still at an early laboratory development stage.

Is it useful to determine the w/c?

The poor precision of the available methods, as well as the limitations on the condition of the concrete that may need to be tested, beg the question whether the w/c of concrete in an exiting structure should be determined in the first place. In considering a situation when the quality of the concrete in a structure is in doubt, the American Concrete Institute (ACI) in its Building Code ACI 318 makes no reference to the determination of w/c, and states that, once concreting has started, the criterion for evaluation and acceptance of the concrete is the compressive strength of test cylinders or cores [5]. Likewise, the Swiss Society of Engineers and Architects states that the w/c can be estimated only on fresh concrete [6].

The various limitations on the condition of the concrete that may be tested (mentioned earlier in connection with the methods of BS 1881: Part 124:1988) are of considerable significance because a decision to determine the w/c of concrete in an existing structure is sometimes taken precisely because it is problematic or damaged, and it is suspected or alleged that this situation has been caused by an incorrect or improper w/c. If the concrete has been damaged, then the w/c cannot be determined so that the question whether too high a w/c is the culprit cannot be answered. In other words, the issue about what is the w/c of the concrete in question becomes converted into the issue of whether the test results are correct.

Part II: Method of optical fluorescence microscopy

In Part I of this section, we have seen that the test methods for the indirect determination of the w/c of hardened concrete that are standardized by British or American organizations do not lead to sufficiently precise results. This does not mean that better test methods should not be sought.

What is the method?

Indeed, there exists one test method, known as Nordtest NT BUILD 361, published in 1991 [7]. As Nordtest methods are little known outside the group of countries that write these methods, some explanation may be in order. The relevant countries are the three Scandinavian countries (Norway, Sweden and Denmark) and Finland and Iceland; collectively, they are referred to as Nordic countries.

I do not know how widespread the use of Test Method NT BUILD 361 is in those countries but, in my experience and to my knowledge, it has not been used in other countries except on an experimental basis. This does not mean that the method is in itself unsatisfactory, but its existence over a decade has not persuaded the international, European or American standard-writing organizations to adopt it or even seriously to consider it for adoption; at least, no evidence to the contrary is available in the literature.

Nevertheless, the Nordtest method deserves a proper discussion. In essence, the determination of the w/c by this method involves

32

impregnating a very small sample of concrete with a low-viscosity resin containing a fluorescent dye, which fills the capillary pores (and other voids). The relative volume of the now coloured voids affects the brightness of the fluorescence in ultraviolet light transmitted through a thin section (about 0.02 mm thick) of the impregnated concrete. The intensity of the colour perceived by eye is compared with that of an appropriate reference sample with a known w/c. The matching colour intensity is taken to represent the w/c of the sample of the concrete being tested.

The practical value of the optical fluorescence microscopy method was studied by St John *et al.* [3]. In their view, the method 'by comparison of the test thin-sections with suitable reference specimens of known w/c, the equivalent w/c of the concrete sample may be estimated to the nearest 0.05 in the range of 0.3 to 0.8' [3]. The need for 'reference specimens' should be carefully noted, and the limitations on these are emphasized by St John *et al.* [3]. St John's own experiments led him to the conclusion that 'Even within the 0.4 to 0.6 range (of w/c) ... values were more realistically gauged to the nearest 0.1, rather than 0.05'. The estimates of w/c were even less reliable at values of w/c below 0.4 and above 0.6 [3].

Another researcher who evaluated the method is Mayfield [8]. In his opinion, the optical fluorescence microscopy method gives a 'qualitative assessment of the w/c of hardened concrete'. The crucial term is 'qualitative'. In an attempt to obtain a quantitative assessment, Mayfield impregnated the concrete using a resin with a dye carried by alcohol. This cannot be used in thin sections and, therefore, he examined the concrete by reflected light, which is distinct from transmitted light in the Nordtest method. I am not aware of further development of Mayfield's approach since the publication of his paper in 1990.

A recent (1999) evaluation of the method of optical fluorescence microscopy was made at Michigan Technological University (MTU) [9]. That work emphasizes the need for reference samples, called standards, which 'should be specifically developed for concretes of different aggregate type, cement type, or curing time'. It is only by comparison of such standards that the w/c of a given sample of concrete can be estimated. Specifically, the researchers at MTU used standards with a range of values of w/c from 0.38 to 0.80, cured for 65 days, and made with aggregate from a local pit, which supplies aggregate for concrete in the relevant area. They emphasize the importance of using the same aggregate in the standards as used in the concrete being tested because 'the green light from the paste surrounding the sand grains causes the sand to glow slightly' [9]; in other words, the colour intensity is affected by the aggregate.

The MTU researchers state: 'The results suggest a trend of increasing brightness with increasing w/c. However the data is rather scattered'. They say that 'the scatter is probably due to the inherent variability within concrete, but might also be due to error introduced during sample preparation' [9]. Indeed their tests on concrete from a highway pavement showed that 50 measurements had a range of values of w/c from 0.30 and 0.66 in one case, and in another, for 58 measurements the range was from 0.49 to 0.93. These numbers give an indication of scatter and, referring to the second set of measurements, they say: 'We do not mean to suggest that a mix design of $0.75\,w/c$ was actually used ... but we do imply that the current condition of the pavement is such that the porosity is equivalent to our 0.75 standard' [9]. My interpretation of this statement is that the porosity could have been induced by deliberate air entrainment or by incomplete compaction, or else during the life of the concrete by agents such as cracking or deleterious expansion. These findings are highly relevant to Part III of this section.

Requirements of the Nordtest method

Despite these discouraging views, the Nordtest method is still being promoted by Danish authors [10], and it is only fair to present the method in some detail. Its scope is 'to estimate the w/c in hardened concrete, using microscopic investigation of thin sections' [7]. The use of the verb 'to estimate' is significant in that the method does not purport to determine the value of w/c, although Section 6.5 of the Nordtest Method NT BUILD 361 uses the expression 'The w/c is determined as the mean value of the w/c in 10 fields of images'.

The precision of the value of w/c determined as the mean value is to be reported to the nearest 0.00 or 0.05. This is significant in that it means that the determination of the w/c is within ±0.05. Moreover, this precision applies only when the cement paste is homogeneous. The Nordtest method states that the range of values of w/c in 10 images is less than 0.1 for homogeneous paste; for a range between 0.1 and 0.2, the paste is considered to be non-homogeneous; and when the range is greater than 0.2, the paste is said to be 'very inhomogeneous'.

What is crucial to the present study of the determination of the w/c of hardened concrete is that the Nordtest method gives a result within ±0.05.

An important limitation of the Nordtest method should be noted: it is applicable to well-hydrated concrete made with Ordinary Portland cement; it follows that other cementitious materials, whose use is nowadays very widespread, must not be included in the mix. Furthermore, there is a requirement for reference samples (standards) that should have the same degree of hydration and the same air void content as the concrete being tested. This is also a very important limitation because, unless proper reference samples are available, it is not possible simply to take a sample of hardened concrete, especially with an unknown curing and exposure history, test it by the Nordtest method, and report the w/c with useful precision.

Some further limitations prescribed in the Nordtest method should be noted. First, the air void content of the cement paste in the field of image in the microscope must be close to that in the reference sample. Second, the cement paste being examined must be uncarbonated and it must have been at least 5 mm distant from the original surface of the concrete.

It seems to me then that the Nordtest method can be successfully used for quality control in production of, say, precast concrete elements. However, with the various limitations given in the last two paragraphs, I would not expect the Nordtest method to be applied to totally unknown concrete and to claim the outcome of the test to give a value of the w/c with a precision of ±0.02. And yet this is what has happened [10]. In my opinion, such a claim should be carefully examined. If it is justified, then a step forward will have been made in the determination of the w/c of hardened concrete. On the other hand, if it is not proved that a precision of ±0.02 is achievable, it is important that those interested in determining the w/c of hardened concrete are not misled. The question of how closely the Nordtest method can determine the w/c of hardened concrete is discussed in Part III of this section.

Part III: A claim of higher precision

The discussion of the determination of the w/c by optical fluorescence microscopy in Part II of this section might lead to the view that the method is, certainly at present, not appropriate for use with unknown concrete and without proper reference samples for comparison of colour intensity. However, in the year 2000, there was published a paper claiming a high precision of such determinations [10].

Such a publication by itself would not merit a serious full-scale rebuttal. However, the 2002 award of the ACI Wason Medal for the

35

paper 'Determination of water–cement ratio in hardened concrete by optical fluorescence microscopy' by Jakobsen *et al.* [10] has brought the matter into focus.

Accuracy of the Nordtest method

The citation for the Wason Medal states that the winning paper describes and *validates* an optical fluorescence microscopy technique to determine the w/c of hardened concrete to an 'accuracy of about ±0.02'. As with all recipients of awards, the authors should be congratulated and nobody would dare to look askance at the choice of the winners because the decision of the powers-that-be is final. The situation is similar to the decision of the House of Lords, which is the highest appeal court of law in the UK. Its decision is final, not because it is right, but it is right because it is final.

It may be useful to cite the definition of the verb 'to validate' given in the *New Shorter Oxford Dictionary*: 'examine for incorrectness or bias; confirm or test suitability'.

What I am writing about is the citation for the Wason Medal, which refers to the validation of 'accuracy of about ±0.02' in the determination of the w/c. This validation is not shown in the body of the paper, but the paper ends with the words: 'Experience with field concrete and with round robin tests performed in Denmark annually show that the expected accuracy of the method is about ±0.02'. I do not know much about experience in Denmark, but the body of the paper presents data that do not support those words. It is the accuracy and precision of the test method that I shall now discuss, leaving the consideration of testing techniques to petrographers.

I can add that, since the award of the Wason Medal for the paper by Jakobsen *et al.*, the Board of Directors of ACI revised the citation for the Medal removing the reference to the *validation* of the 'accuracy of about ±0.02'. The revised citation for the Wason Medal, published in *Concrete International*, June 2003, now reads: 'For their paper describing an optical fluorescence microscopy technique for estimating the water–cement ratio of hardened concrete'.

Views of other researchers

I cannot help noticing that all but one of the references to the use of the technique to determine the w/c are to Danish sources, for the most part including the authors of the paper that I am discussing [10]. Failure to discuss, and possibly to refute, the views of St John *et al.* [3] is surprising. The absence of references to the use of the fluorescence method in countries other than Denmark may possibly be explained by the non-acceptance of the method in the rest of the world, or else by a selective choice of references. In this connection, in *Neville on Concrete* I discussed 'objectivity in citing references' [11].

An international reference mentioned in the paper by Jakobsen *et al.* [10], is a paper by Elsen *et al.* [12] on the use of optical fluorescence microscopy in conjunction with image analysis. That paper reports tests on cement paste and concrete samples cured under laboratory conditions for 28 days, and says: 'The results cannot of course be transferred in a simple way to field concrete' [12]. I do not see the preceding statement as supporting the thesis and conclusions of Jakobsen *et al.* [10].

What is meant by accuracy?

The paper by Jakobsen *et al.* claims 'an accuracy of about 0.02' [10]. What exactly is meant by this? In common parlance, accuracy means that your rifle shot hit the bull's eye, or very nearly so. If you fired a number of shots whose average was the bull's eye, then the term accurate would be appropriate. So, accuracy is to do with being near the 'true' value. On the other hand, if you manage every time to hit the same spot, but that spot is away from the bull's eye – say, in military parlance, it is in the 'outer' – then your performance is not accurate, although you achieve a high precision.

I would like to illustrate these concepts by quoting from Kennedy and Neville [13]: '*Accuracy* is the closeness or nearness of the measurements to the "true" or "actual" value of the quantity being measured. The term *precision* (or repeatability) refers to the closeness with which the measurements agree with each other.' The distinction between accuracy and precision is illustrated in Fig. 2.2.1 in this section.

For the purpose of considering testing specimens of hardened concrete, we should use ASTM E 177-90a *Standard Practice For the Use of Terms Precision and Bias in ASTM Test Methods*. This *Standard Practice* has changed the meaning of the term accuracy from earlier editions, and it now says that 'No formula for combining the precision and the bias of a test method into a single numerical value of accuracy is likely to be useful'. The term *bias* of a measurement process is said to be 'a generic concept related to a consistent or systematic difference between a set of test results from the process and an accepted reference

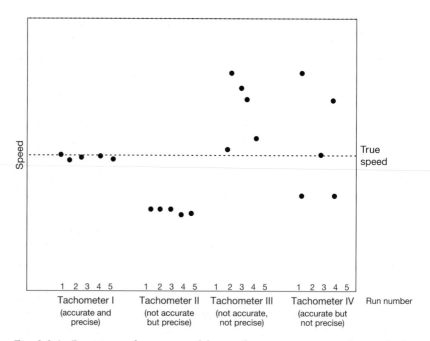

Fig. 2.2.1. Precision and accuracy of four tachometers measuring the speed of a constant-speed motor [13]

value of the property being measured'. Significantly, ASTM E 177 says that 'when an accepted reference value is not available, the bias cannot be established'.

The preceding statement is of importance in relation to the method described by Jakobsen *et al.* [10] in tests on concrete in cores from concrete elements several years old. In those cases, there exist no reference values because nobody now knows what was the true w/c in the concrete at the time of placing. We know what was specified, and we can assume that the ready-mixed concrete supplier took appropriate steps to feed into the mixer the appropriate proportions of the various mix ingredients. But we also know that, inevitably, there are errors in the batching process, especially with respect to the mass of water on the surface of aggregate (or deficient in the pores of aggregate particles in a condition drier than that described as saturated-and-surface-dry). Furthermore, evaporation of water from concrete in transit can affect the w/c at the time of placing of concrete.

There is one more important reason why concrete in a structure does not have the same w/c throughout. There is a variability in the concrete from place to place, both in the horizontal direction, owing to placing and consolidation procedures, and in the vertical direction, owing to bleeding. It is important to note that the specimens used in optical fluorescence microscopy are very small compared with structural elements: 35 by 45 mm in size and 0.02 to 0.03 mm thick.

For all these reasons, there is no reference concrete and therefore no accepted reference value of the water–cement ratio. In my opinion, therefore, however precise the test method when applied to a single piece of hardened concrete that was subdivided into small samples tested by various laboratories or operators, no statement about bias is possible. In other words, it is not possible to talk about the accuracy of the test method for the purpose of establishing the w/c of unknown old concrete. Moreover, the exposure history of a given concrete, which affects the degree of hydration of cement, influences the volume of capillary pores liable to be filled by the resin: for the same w/c, the smaller the degree of hydration the larger the volume of capillary pores; this results in a spuriously high estimate of the w/c.

In this connection, it is useful to recall that ASTM E 177 in its 1990a edition, abandoned the term 'accuracy' and says that 'only the terms precision and bias should be used as descriptors of ASTM test methods'. Admittedly, the Nordtest method is not used in the USA or in Europe, with the exception of Nordic countries but, for the benefit of world-wide readers, adhering to the rigorous approach of ASTM is salutary.

What is meant by precision?

Let me now turn to the consideration of precision. ASTM E 177 describes precision as 'a generic concept related to the closeness of agreement between test results obtained under prescribed like conditions from the measurement process being evaluated'. It is said that precision has no meaning unless the measurement process is in a state of statistical control, which means that variations between the observed test results can be attributed to a constant system of chance causes. Chance causes are unknown factors, each small but numerous, and not readily detectable or identifiable.

In this connection, it is important to note that, according to ASTM E 177, the presence of outliers may be evidence of a lack of statistical control. This is of relevance to the paper by Jakobsen *et al.*, who give the impression of considering outliers as 'bad results' that should be ignored without an assigned cause, and they applied that approach in the interlaboratory (round robin) test [10].

Range of water–cement ratios tested

The paper by Jakobsen *et al.* states that the visual determination of the w/c is made 'in the range of 0.35 to 0.70' and that two different operators can determine 'concrete with known w/c by an accuracy of ±0.02 within the target value' [10]. The paper reports the determination of w/c by two operators on concrete with a w/c of 0.35 only; no results for higher values of w/c are given. Yet, in practical situations, higher values of w/c are often used and may well require verification.

Furthermore, it is useful to consider the *rate* of change in capillary porosity with a change in the w/c. According to a table in their paper, the change in porosity between mixes with a w/c of 0.40 and of 0.45 is 6 percentage points, whereas between the values of w/c of 0.65 and of 0.70, it is 3 percentage points. In my opinion, in a general way, the rate in the change of capillary porosity is very much higher at low values of w/c than at high values.

The significance of this situation is that the sensitivity to the intensity of the colour change in the impregnated concrete is bound to be lower at high values of w/c so that the bias (which, I think, is what Jakobsen *et al.* are reporting) is likely to be lower at high values of

Table 2.2.1. Visual determination of w/c performed by two different operators (reproduced from [10])

Thin section no.	Operator 1	Operator 2	Average	Difference 1–2
1.1	0.35	0.35	0.35	0.00
1.2	0.35	0.37	0.36	−0.02
1.3	0.38	0.37	0.38	0.01
2.1	0.33	0.32	0.33	0.01
2.2	0.34	0.35	0.35	−0.01
3.1	0.34	0.35	0.35	−0.01
3.2	0.32	0.32	0.32	0.00
4.1	0.35	0.35	0.35	0.00
4.2	0.35	0.35	0.35	0.00
4.3	0.36	0.35	0.36	0.01
5.1	0.35	0.33	0.34	0.02
5.2	0.35	0.35	0.35	0.00
6.1	0.37	0.40	0.39	−0.03
6.2	0.36	0.40	0.38	−0.04
7.1	0.35	0.37	0.36	−0.02
7.2	0.35	0.37	0.36	−0.02
Average	0.350	0.356	0.353	−0.006
SD	0.014	0.023	0.018	0.016

Note: SD = standard deviation

w/c than found in their tests by two operators on concrete with w/c of 0.35 (see Table 2.2.1). This fact is recognized in the body of their paper, but the necessary limitation on w/c is not included in the conclusions of the paper, which is what many readers are likely to use as the first source of information.

Inter-operator tests

Looking at Table 2.2.1, we should ask: what is the precision of their data? One operator's testing resulted in a standard deviation, s, of 0.014, the others, 0.023; the table reports an average s of 0.018. The authors do not explain the term 'average' but 0.0185 is the arithmetic mean of the values of the standard deviations of the two operators. Standard deviations should not be averaged arithmetically: the mean value of the standard deviation is the root-mean-square value, that is, the square root of one-half of the sum of squares of the individual values. Even that is an approximation because the number of degrees of freedom is one less than the sample size (i.e. 2). The properly calculated value of the mean standard deviation is 0.019. In this paper, this value will be used henceforth to establish the confidence interval of the mean value of w/c as determined by the two operators in Table 2.2.1.

The confidence interval commonly used in testing is that within which 95% of test results are likely to fall. ASTM E 77 uses the 'two standard deviation limits' although the strict index is $1.96s$ rather than $2s$. The ASTM definition of these limits is: 'Approximately 95% of individual test results from laboratories similar to those in an interlaboratory study can be expected to differ in absolute value from their average value by less than $1.960s$ (about $2.0s$).' Incidentally, the standard deviation should be determined on at least 30 test results; in the paper being discussed, the number of test results was 16 [10].

Now 2.0 standard deviations is 2×0.019, calculated above, that is 0.038 or 0.04. It follows that, if the true value of w/c was 0.035, then 95% of results should fall within the range of 0.031 to 0.039. Looking at Operator 1, we find that all 16 values fall within this range. For Operator 2, two out of 16 fall outside the range. This means that 88% are within the range, which does not necessarily contravene the 95% limits, given the small number of tests.

However, if the claim is that ± 0.02 is the confidence interval, then we should look at results outside the range 0.33 to 0.37. For Operator 1, two test results fall outside and for Operator 2, three test results. In addition, there are several marginal values of 0.32 and 0.37. Thus the claim of determining the value of w/c within ± 0.02 clearly fails. It fails also on the basis of the values of the standard deviation reported in Table 2.2.1: twice the standard deviation for each of the two operators is, respectively, 0.03 and 0.05.

Interlaboratory tests

The results of interlaboratory tests (also called round robin tests) are partially reported in the paper by Jakobsen *et al.*, where it is said that 'Table 5 shows some results of the round robin testing' [10]. That table is reproduced as Table 2.2.2. The claim that 'The round robin test is by itself a powerful tool to keep up the standards of optical microscopy amongst the laboratories' is valid, but I fail to see the validity of the conclusion that 'Experience with field concrete and with round robin tests performed in Denmark annually show that the expected accuracy of the method is about ± 0.02' [10].

First, the purpose of interlaboratory tests is to assess the performance of the laboratories involved. The ASTM E 691-87 *Standard Practice for Conducting an Interlaboratory Study to Determine the Precision of a Test Method* is clearly relevant and also quite demanding.

If anything, the interlaboratory testing, as described in Table 2.2.2, gives information on the precision of testing but not on the bias of the results. Indeed, the *input* w/c, that is, the w/c of the concrete as batched is given in only four out of the seven annual test rounds. It is inconceivable that the input value was not known to the test organizers

Table 2.2.2. Round robin results of determination of w/c (reproduced from [10])

Concrete Year	A 1987 – I	B 1987 – II	C 1990	D 1991	E 1992	F 1993	G 1994
Laboratory 1	0.40	0.35	0.40	0.38	0.49	0.40	0.40
Laboratory 2	0.40	0.35	0.35	0.40	0.55	0.38	0.40
Laboratory 3	0.45	0.40	0.35	0.40	0.45	0.40	0.40
Laboratory 4	0.40	0.35	0.40	0.50	0.45	0.35	0.47
Laboratory 5	–	–	0.40	0.40	0.40	0.43	0.37
Laboratory 6	–	–	0.40	–	0.50	–	–
Laboratory 7	–	–	0.35	–	–	–	–
Mean	0.41	0.36	0.38	0.42	0.47	0.39	0.41
SD	0.025	0.025	0.027	0.048	0.052	0.029	0.037
w/c from mix	–	–	0.37	0.43	0.44	0.42	–

Notes: SD = standard deviation

– means that data were not obtained or were not available

in all the cases and, yet, Table 2.2.1 describes the other cases as 'data not obtained or not available'. Moreover, in 2 out of 4 test rounds in which the input w/c (that is, the value as batched) is given, the difference between the input value and the mean value of the test results is 0.03 (positive in one case, negative in the other). How, then, can it be claimed that the accuracy is 'about ±0.02'?

It may be useful to make a further reference to ASTM Practice E 691. This states that the 'practice is designed only to estimate the precision of a test method', and further on, 'when accepted reference values are available for the property levels, the test result data . . . may be used in estimating the bias of the test method'. No reference values are reported in the paper by Jakobsen *et al.* Furthermore, ASTM E 691 states in bold letters: 'Under no circumstances should the final statement of precision of a test method be based on acceptance of test results for each material from fewer than 6 laboratories. This would require that the inter-laboratory study begin with 8 or more laboratories in order to allow for attrition.' Admittedly, it is possible that in a country the size of Denmark there are not enough laboratories to satisfy the ASTM requirement but, if so, the test method should not be 'exported' to the European Union or to the USA.

A curious feature of the interlaboratory test results of Jakobsen *et al.* is that in three years they are reported to the nearest 0.05; in two further years, all but one result is given to the nearest 0.05. This approach is inconsistent with the claim that the *accuracy* is ±0.02. Moreover, although seven laboratories were involved, only in one year are the results of all seven reported; it would be interesting to know whether the other laboratories obtained 'bad' results or just failed to conduct the tests in a proper manner.

As far as the quality of the results reported by the various laboratories is concerned, Jakobsen *et al.* refer to 'Examples of laboratories out of line' [10]. In one case, a particular laboratory reported the value of w/c of 0.50 while the input value was 0.43, and the mean for the parti-cipating laboratories was 0.42; in the second case, the reported value was 0.55 while the input value was 0.44, and the mean value for the two laboratories was 0.47. These are very large deviations: 0.07 and 0.11 from the 'true', that is, input value, and a far cry from ±0.02.

To say that the laboratory is 'bad' is to acknowledge that the test method is not 'rugged' in the terminology of ASTM E 177, or that the level of competence required in the subjective assessment of the intensity of fluorescence, which is the 'measure' of the w/c, is not achievable by some laboratory testers. The ruggedness tests are prescribed by ASTM E 1169-89. These tests should precede the inter-laboratory tests, their purpose being to study the test method variables and to indicate the need for selective tightening of test method specifications, and hence to ensure appropriate precision of the test method.

With respect to the variability between the laboratories, appropriate calculation methods are given in ASTM E 619. Unfortunately, the interlaboratory tests reported by Jakobsen *et al.* [10] appear to be neither adequately organized nor fully presented. However, what is apparent for the four laboratories that tested the concrete in all the seven years, is that the average w/c for each laboratory is very nearly the same: 0.403, 0.404, 0.407 and 0.417. In other words, every year the concrete tested had the same input w/c. On the other hand, in any one year, the average value of the w/c for the four laboratories differed considerably: the values for the seven concretes A to G are: 0.412, 0.357, 0.375, 0.420, 0.485, 0.382 and 0.417. Does this mean a large between-laboratories variance? Analysis of variance would answer that question.

In this connection, we should note that ASTM E 691 uses values of 95% limits on repeatability and reproducibility and gives expressions for the standard deviation (that is, the square root of the variance) of each of these two parameters. The repeatability and reproducibility limits are 2.8 times the repeatability and reproducibility standard deviation, respectively.

Tests on field concrete

Jakobsen *et al.* reported the use of their test method in quality control testing of concrete used in the construction of Copenhagen airport [10]. It is for such a purpose that optical fluorescence microscopy is likely to be suitable, using a single 'experienced petrographer' with well-established reference samples.

The situation is, however, not the same in the case of concrete such as railway ties that 'were in field service for a number of years before determination of w/c' [10]. Here, too, 'a single experienced operator did the determinations'. The average value of w/c was 0.42, the standard deviation was 0.026, the number of ties tested was 127 and the number of test results was 522. If the accuracy is, as claimed, ±0.02, then 2.14% of 522, that is approximately 11 test results, should fall below $0.42 - 0.02 = 0.40$, and 2.14% above $0.42 + 0.02 = 0.44$. As reported in this chapter, the number of test results below $w/c = 0.42$ is about 145, and the number above 0.44 is about 100 (see Fig. 2.2.2;

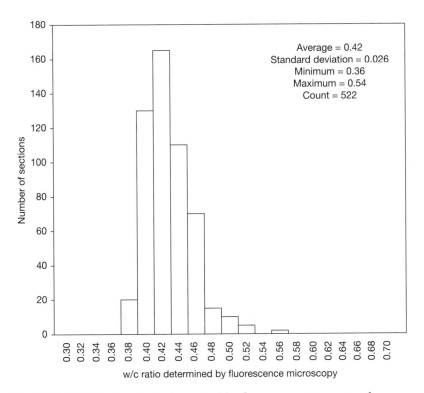

Fig. 2.2.2. Distribution of w/c determined by fluorescence microscopy of concrete railroad ties. W/c of mix design was 0.44. Data from G.M. Idorn Consult, RAMBØLL, 1992 (reproduced from [10])

the inconsistencies in the figure are in the original paper). These values are not surprising because the standard deviation reported by Jakobsen et al. [10] is 0.026. Hence, the correct 95% confidence limits are $0.42 - 2 \times 0.026 = 0.37$ and $0.42 + 2 \times 0.026 = 0.47$. In other words, the optical fluorescence microscopy tests have shown that the actual w/c of the concrete in the ties lies between 0.37 and 0.47. This value of 0.1 is exactly the range for determination of w/c of *unknown concrete* using existing methods of estimating the w/c given in *Properties of Concrete* [14]. As I see it, the superiority of the fluorescent microscopy tests is illusory, at least at present.

Conclusions

A quantitative study of the determination of w/c of hardened concrete by optical fluorescent microscopy, as reported by Jakobsen et al. [10]

does not support the claim that in field concrete 'the expected accuracy of the method is about ±0.02'. My findings accord with the views of St John *et al.* [3] and they give a helpful background to the fact that the method is not used by ASTM or in European Standards. Indeed, what the data of Jakobsen *et al.* demonstrate is that w/c of an unknown concrete can be determined within 0.1. This can be achieved by other techniques as well. For practical purposes, a value of w/c within 0.1 is not of much use, and certainly not of probative value in disputes where large sums of money can depend upon the proof of a compliant value of w/c.

It is worth recalling that St John *et al.* found fluorescence microscopy to be 'particularly useful for detecting small-scale variations in w/c' [3]. This is what is being controlled in repetitive production of concrete.

It seems, therefore, that the technique described by Jakobsen *et al.* may serve as a tool in quality control in the production, for example of tunnel lining segments using a single operator, but not to determine the w/c of hardened concrete in a reliable manner, especially in the absence of contemporaneous reference samples made with the same materials and with the same exposure history.

As often is the case, future improvements in testing methods are possible but, for the time being, the case for superiority of optical fluorescence microscopy with an accuracy of ±0.02 lacks statistical support. We can, therefore, remain in the belief that a reliable and precise determination of the w/c of a sample of hardened concrete taken from a structure with an unknown history of curing and exposure, especially, when damage or deterioration is suspected, is at present not available. An estimate within ±0.05 is the best that we can do.

References

1. Neville, A., How useful is the water–cement ratio?, *Concrete International*, 21, 9, 1999, pp. 69–70.
2. Analysis of hardened concrete, *Technical Report 32*, The Concrete Society, 1989.
3. St John, D.A., Poole, A.W. and Sims, I., *Concrete Petrography*, Arnold and John Wiley, London and New York, 1998.
4. Bate, S.C.C., *Report on the failure of roof beams at Sir John Cass's Foundation and Red Coat Church of England Secondary School, Stepney*, Building Research Establishment, 1974, Current Paper 58.
5. *Building Code Requirements for Structural Concrete*, ACI 318, American Concrete Institute, 1999.

6. Définition et détermination du rapport eau sur ciment, *Societé Suisse des Ingénieurs et des Architects*, 7, April, 1999, pp. 128–129.

7. Concrete, hardened: water–cement ratio, *Nordtest Method*, *NT Build*, 361, Nordtest 1991-02.

8. Mayfield, B., The quantitative evaluation of the water/cement ratio using fluorescence microscopy, *Magazine of Concrete Research*, 42, 150, 1990, pp. 45–49.

9. *A real world application of the MTU w/c standards*, Michigan Technological University at www.cee.mtu.edu/html (March, 2003).

10. Jakobsen, U.H., Laugesen, P. and Thaulow, N., Determination of water–cement ratio in hardened concrete by optical fluorescence microscopy, *Water–cement Ratio and Other Durability Parameters – Techniques for Determination*, M.S. Khan, Editor (SP-191, ACI, Michigan, 2000) pp. 27–41.

11. Neville, A., Objectivity in citing references, *Neville on Concrete*, Section 8-1, ACI International, Farmington Hills, MI, 2003.

12. Elsen, J., Lens, N., Aare, T., Quenard, D. and Smolej, V., Determination of the w/c ratio of hardened cement paste and concrete samples on thin sections using automated image analysis techniques, *Cement and Concrete Research*, 25, 4, 1995, pp. 827–834.

13. Kennedy, J.B. and Neville, A.M., *Basic Statistical Methods for Engineers and Scientists*, Third Edition, Harper and Row, New York, 1986.

14. Neville, A.M., *Properties of Concrete*, Fourth Edition, Longman and John Wiley, London and New York, 1995.

2.3 COURT DECISION ON THE USE OF THE OPTICAL FLUORESCENCE MICROSCOPY TEST

Judge Hon. David C. Velasquez in Superior Court of California, County of Orange, 7 June 2005

Defendants applied to court to exclude evidence and testimony of the determination of the water–cement ratio based on the use of optical fluorescence microscopy, discussed in Section 2.2. The primary grounds for exclusion is that the method lacks general acceptance in the relevant scientific community. Below is the judge's ruling, given *verbatim*.

Water–cement ratio

Defendants' motion to exclude evidence and testimony of the water to cement ratio in the original mix design of eight year old in-service concrete by petrographic means is granted.

The use of petrographic means to determine the water to cement ratio of the original mix of hardened concrete is generally accepted for purposes of research, where known reference samples have been prepared, or when in-service concrete is relatively young. (...) However, '[b]ecause of the variable nature of cement pastes, the age of pastes, and exposure to a variety of external influences, there is no generally accepted standard procedure that employs microscopical methods for determining the w/c or w/cm of hardened concrete...' which has been in service for the number of years as the concrete involved in the instant action (ASTM 856 §12.3). While estimation of the original w/c of hardened concrete by petrographic means is useful for purposes of quality control in research, where comparisons can be made to known reference samples, such estimations cannot be reliably made for concrete which has been in place for a number of years under uncontrolled circumstances. The aging process of concrete is very complex. The estimation of the original w/c must contend with many variable influences such as the characteristics of the mixing and placing processes, vagaries in the sampling methods, variable temperature and moisture, the presence of admixtures, the complexity of the hydration process, re-mixing of the concrete at the site of placement, questions about the source of the cement used to make the paste, alterations of the physical properties of the paste due to chemical alteration such as carbonation, changes in porosity due to leaching, and the subjective abilities of the petrographer.

Plaintiffs have offered several tests approved by European industrial standards to determine the original w/c of hardened concrete. Plaintiffs argue that such methods have also been used in Canada and the United State, *ergo*, they are generally accepted in the relevant scientific community. However, plaintiffs have not presented persuasive evidence that scientists and researchers here in the United States have generally accepted such methods. Neither side has presented the court with authority of what constitutes a relevant scientific community. Plaintiffs urge the court to find that the relevant scientific community in the instant case is global in scope. This may be the case given the international dialogue and involvement of the scientists on both sides of the Atlantic Ocean concerned with the investigation, research, design and study of the physical properties of concrete. But assuming that to be the case, plaintiffs have not shown, by a preponderance of evidence, a consensus of a cross-section of the world-wide scientific community on the tests at issue herein to determine the w/c of ten year old field concrete.

The court finds that there are industry standards in place in Europe supportive of some of the methods used by plaintiffs' experts to determine the water–cement ratio of relatively old field concrete. However, plaintiffs have not carried their burden to show the European standards represent an *international* consensus. The fact the European standards are not accepted by those involved in setting consensus standards here in the United States is evidence of the lack of an international consensus on those test methods proffered by the plaintiffs. Such disagreement by major voices in the scientific community on this side of the Atlantic is strong evidence that such tests are not generally accepted either here in the United States or internationally.

3

High-alumina cement

High-alumina cement is not a type of Portland cement, or even a blend of Portland cement, with which the major part of this book is concerned. Rather, high-alumina cement is a specialist cement, made of raw materials different from those used to manufacture Portland cement, employing a different process of manufacture and used for specific purposes. One could, therefore, ask why this chapter is devoted to high-alumina cement.

The answer is that in the past, up to about 30 years ago, high-alumina cement was used in a number of countries, including the UK and Spain, in the manufacture of prestressed precast concrete units. This was to exploit the very high, early strength of concrete made with high-alumina cement, which allowed rapid removal of formwork and its re-use in a 24-hour cycle of manufacture of standard precast concrete units. Alas, the products of hydration of high-alumina cement are chemically unstable and undergo a change known as conversion, the change being more rapid the higher the temperature. Conversion results in an increase in the density of the cement hydrates, with a consequent increase in porosity and a reduction in strength.

The loss in strength of the concrete is significant and reduces the safety of the structure; in some cases, collapse has followed. For this reason, high-alumina cement was deleted from the list of cements permitted in reinforced and prestressed concrete, and thus effectively banned from routine use, not only in the UK but also in a great many other countries; the only significant exception is France. In some other countries, standards and codes make no reference to high-alumina cement because the cement is not produced there, and its price is much higher than Portland cement, so that there has been no incentive to use it.

The preceding 'history' of the structural use of high-alumina cement concrete still does not explain the inclusion of the discussion of that cement in a 2006 book. Thus, a justification is called for. Briefly, in the last few years, there has been a concerted effort by the manufacturers of high-alumina cement to re-introduce the structural use of that cement. One means of doing so was to establish a working party, composed mainly of cement chemists, which reviewed the past experience and recommended a 'new look'. Another means was to use a new European standard for high-alumina cement – a part of the series of European standards for cements to include a large annex on structural use of high-alumina cement.

I view these developments as dangerous and I discuss them in Sections 3.1 and 3.2. Section 3.3 reviews in detail the behaviour of concrete made with high-alumina cement in structures. The ensuing discussion at the meeting of the Institution of Structural Engineers, at which I presented my paper, shows that mine is not a voice crying in the wilderness.

I commend a perusal of Chapter 3 as a salutary tale of the need for wariness in accepting new, or newly 'dusted off', materials in the construction industry. The old Roman maxim 'Buyer, beware!' is still valid.

The defence against the reintroduction of high-alumina cement into structural use must continue. An example of the continuing onslaught is shown in advertisements by Lafarge cement manufacturers in a Mexican structural engineering journal recommending the use of high-alumina cement (Ciment Fondu Lafarge) in highways, runways, tunnels, jetties, slabs and foundations.

3.1 DRAFT STANDARD FOR HIGH-ALUMINA CEMENT: SHOULD IT TELL US HOW TO MAKE CONCRETE?

A Draft British Standard, BS EN 14647:2003, *Calcium Aluminate Cement – Composition, Specifications and Conformity Criteria* [1], has been put out for final comment. If the document is published as a Standard, it will supersede the current BS 915-2:1972 *Specification for High Alumina Cement – Metric Units* [2]. Normally, a new Standard for cement would not be of broad interest to engineers but, in this case, there are several unusual circumstances.

The importance of the proposed standard

The main unusual feature of this Draft Standard is that it contains a major annex on the use of the cement in concrete and it is this that causes concern. Approval in the UK is not a necessary condition for the Standard to become mandatory in this country. The introduction to the Draft British Standard states that, even if not approved in this country, 'the UK will be obliged to publish the official English language text unchanged as a British Standard and withdraw any conflicting Standard' if the necessary support elsewhere in Europe is obtained. Therefore, the acceptance of the Standard in Europe will determine what happens in this country.

The name of the cement

Apart from the use of proprietary names, the cement has been known in the UK as high-alumina cement (HAC) and this name is used in BS 915-2:1972, including its latest amendment dated 8 April 2003. The Draft Standard, however, uses the new name of calcium aluminate cement and describes 'high-alumina cement' as a previous name. Yet, none of the publications cited in the bibliography at the end of the Standard uses the name 'calcium aluminate cement'. The Building Research Establishment has used the name 'high-alumina cement' as recently as 2002. In English, the acronym HAC is well established, and it is a matter for conjecture whether the change is to distance 'calcium aluminate cement' from the old HAC associated with the structural failures in the 1970s and its subsequent removal from the structural codes.

A recent re-assessment of use of HAC in construction was prepared by The Concrete Society [3] and is discussed in my book *Neville on Concrete* [4].

The annex

The Draft Standard contains an annex of nine pages, which is described as informative. This gives the impression that it may do no harm, regardless of whether it is right or wrong; however, given the topic it covers, this may be an illusion.

The title of the annex is 'Essential principles for use of calcium aluminate cement in concrete and mortar'. However, what follows is not 'essential principles for use' but rather descriptive material. The use of cement in concrete seems an inappropriate topic for inclusion in a Standard for cement. European Standards for other cements contain no annex on their use in concrete. Such use is the province of concrete Standards and Codes of Practice; for example, BS 8110-1: 1997 *Structural Use of Concrete – Code of Practice for Design and Construction* [5].

Terminology, and selective and missing advice

The use of unambiguous language and precision in the choice of words are vital in a Standard. In the new Draft Standard there are many exceptions to this requirement. Some terms used are not those commonly used in engineering. Good resistance to temperature and to corrosion are unspecific claims. To say that 'concreting in hot weather can be carried out with no risk...if certain precautions are taken' is a bold claim. It is said that 'the time to achieve conversion is defined by the time to reach minimum strength', and yet Collins and Gutt [6], who are cited in the Bibliography of the Standard, have shown that under wet conditions there is a continuing loss of strength beyond full conversion; this is important.

The annex is selective and refers only to a few aspects of making concrete. Thus, a false impression could be created that only the aspects considered in the annex are important. Alternatively, it could be thought that the 'essential principles' deal with those that are different from those applicable to concrete made with Portland cement. However, this is not so, as exemplified by exhortations to clean the formwork and to cure with water or curing compounds, as well as to compact using vibrating pokers, all of which apply when Portland cement is used.

Some specific features of concrete made with HAC are not mentioned, especially regarding durability. HAC concrete is described as 'resistant to chemical attack' while HAC concrete after conversion is especially vulnerable to alkalis carried by percolating water from

Portland cement screed or lime. Carbonation and consequent corrosion combined with alkaline hydrolysis are not mentioned.

It is likely that the Draft Standard was not written by engineers. It is perfectly reasonable for a Standard for cement to be written by chemists, but an annex dealing with the use of cement in concrete is the province of engineers. The names of the members of the committee that wrote the Draft Standard are not divulged by British Standards Institution, and this does not dispel my suspicion that many of the chemists are representatives of the two sole manufacturers of HAC clinker in Europe.

Bibliography
In addition to Standards, there are four publications listed, the most recent of which was published in 1988. Of these, three are by chemists, two of whom worked for a cement manufacturer. The only paper written by an engineer was written in 1975. I may be egocentric, but I feel that my 200-page book *High-alumina Cement Concrete* [7], published in the same year, could have been cited. More significantly, numerous papers published since then are ignored.

Concluding remarks
This Section may give the impression of a catalogue of complaints. If so, my justification is that if an inadequate Standard comes into force, it will be extremely difficult to change it by the will of the people in this country. This aspect of European Standards may not be fully appreciated. With respect to HAC, the situation is particularly delicate because the use of HAC in concrete, other than in refractory concrete, with which the annex is not concerned, is limited to a very small number of countries. It is, therefore, of little consequence across much of Europe that the Standard is technically correct. So now is the time to be concerned about the inclusion of the annex in Draft BS EN 14647:2003 [1].

References
1. British Standards Institution, BS EN 14647:2003, *Calcium Aluminate Cement – Composition, Specifications and Conformity Criteria*, London.
2. British Standards Institution, BS 915-2:1972, *Specification for High Alumina Cement – Metric Units*, London.

3. The Concrete Society, Technical Report 46: *Calcium aluminate cements in construction – a re-assessment*, The Concrete Society, 1997.
4. Neville, A., *Neville on Concrete*, American Concrete Institute, Farmington Hills, MI, 2003.
5. British Standards Institution, BS 8110-1:1997, *Structural Use of Concrete – Code of Practice for Design and Construction*, London.
6. Collins, R.J. and Gutt, W., Research on long-term properties of high-alumina cement concrete, *Magazine of Concrete Research*, 40, 145, Dec., 1988, pp. 195–208.
7. Neville, A. and Wainwright, P., *High-alumina Cement Concrete*, The Construction Press, Lancaster, 1975.

3.2 REVISED GUIDANCE ON STRUCTURAL USE OF HIGH-ALUMINA CEMENT

In the July/August 2003 issue of *Concrete* there was published my article [1] [Section 3.1] on a Draft British Standard BS EN 14647:2003 *Calcium Aluminate Cement – Composition, Specifications and Conformity Criteria* [2]. Calcium aluminate cement is a new name introduced by the manufacturers for high-alumina cement (HAC) which acquired notoriety in the structural roof failures in the 1970s. In that article, I pointed out certain flaws in terminology and, more importantly, I criticized the inclusion of an annex on the use of HAC in *structures*.

Composition of the Standard Committee

As my article criticized BS 14647:2003, I expected some reaction by way of letters: there was none. I am somewhat surprised by the silence of the manufacturers of HAC or of those involved in the preparation of the standard. I am, alas, old enough to have learned Latin, and I remember the tag to the effect that those who remain silent appear to agree. What seriously worries me is that the members of the Committee remain under the cover of anonymity. What I was interested in was not the names of members but their profession and their employers. My enquiries were treated with courtesy, but they elicited no information about the composition of the Committee, which wrote a standard with an annex promoting the structural use of HAC.

It is said that justice has not only to be done, but also has to be seen to be done. When I was training to become an arbitrator, I was told that it is not good enough for an arbitrator to be independent: he must not fail to disclose that he owns shares in, for example, a subsidiary company that is a party to the arbitration. Of course, there has to be a sense of proportion in all that, and disclosure to the parties may be accepted as not prejudicing independence. Anyway, a judge whose independence was once challenged by the fact that his wife owned 20 shares in Shell (involved in the case) said: 'I know that every man has his price but I am offended by your thinking that mine is so low.'

By way of contrast, in my experience, when I was a member of the committee that drafted the structural codes BS CP110 and 8110, all the names of members were listed in the codes. I am told that this practice is no longer followed in order to protect the committee members from being pursued in the courts for alleged loss caused by the application of the Standard. Such secrecy is fraught with danger, and is not acceptable in other spheres. We are told that there should be more open government. The Bank of England committee that fixes the bank rate publishes how each member of the committee has voted, and even gives the member's reason for the vote. Judges are vulnerable to revenge by those whom they have sentenced, but they do not wear hoods in court. Standards are custodians of appropriate quality of materials and correct construction methods and procedures. But, if the custodians are anonymous, they are not seen to be responsible to anyone and to be not influenced by occult factors.

Revised Standard

In response to the call by BSI for comments on Draft BS EN 14647:2003, I sent some comments to the British Standards Institution, and no doubt so did many other people. These comments led to a significant revision of the Draft Standard, which may become the final version. Let us then look at the modified document.

As far as the body of the Standard is concerned, the changes were minimal, but I view it as unfortunate that Section 1 of the Standard, headed 'Scope', has been expanded by the addition of the statement: 'Principles for the correct use of calcium aluminate cements (HAC) are given in Annex A'. This could be interpreted to mean that the structural use of concrete containing HAC is routine, which of course it is not.

Revised annex on structural use

The Introduction recognizes now that HAC is not an obsolete name, but is still an acceptable alternative. Incidentally, the old German name, *Tonerdezement* has remained unaltered.

My comment submitted to the drafting committee, to the effect that the annex on the structural use of concrete does not fit into a standard for cement, was not accepted. However, the title of the annex was changed from 'Essential principles', which has a normative mandatory nature, to 'Guidance for the use'. This is still to be confirmed. Moreover, a number of positive changes to the annex have been made.

In the Introduction of the Annex A.1, it is explicitly stated that 'it is essential to take into account the conversion phenomenon'. Moreover, in the section on specific characteristics of the cement, it is said that 'conversion is inevitable and irreversible' and also: 'Complete conversion may take several years at 20 °C . . .'. Such a view of conversion is correct, but it is a radical change from the views forcefully expressed in 1964 in the discussion of my paper to the Institution of Civil Engineers [3]. For example, in §157 of [3] J.K. Sykes (a contractor) said of my opinion 'that conversion was inevitable, and a loss of strength a direct consequence, this was not the case in practice. In Great Britain, at any rate, conversion rarely occurred [never in Mr Syke's experience].'

Likewise, in §124 [3], O.J. Masterman (a prestressed concrete manufacturer) 'took exception to the statement in §37 of the paper by Neville that at ordinary temperatures prevailing in England conversion took place spontaneously'. And further on, in §126, Masterman said: 'One need not worry about high-alumina cement which was properly made and used in suitable conditions.'

Still in the 'Introduction' of the Annex A.1, sweeping but unspecific statements about special properties of HAC, such as 'resistance to temperature, to wear and corrosion', have been removed. This is welcome.

In Section A.3.4 of the Annex on 'Strength development', it is recognized explicitly that 'Over time, with the progression of the conversion phenomenon, strength will decrease to a minimum stable level . . .'. It is also said: 'For design purpose, only minimum strength level after conversion must be considered.' This is a correct way of considering the strength of HAC concrete, especially as it is also said in the same section 'that converted strength cannot be estimated from the unconverted strength, since the ratio between the two is not constant'. The need to consider explicitly the strength after conversion was denied in the past, yet it is very important for the purpose of conforming to the 'Approved Document' to the 1999 Amendment to Building Regulations, which requires proof that the 'final residual properties (of materials), including their structural properties, can be estimated at the time of their incorporation into the work . . .' [4]. It is useful to recall that designing on the basis of future minimum strength was not approved by some designers. For example, in §183 of [3], T.N.W. Akroyd (a consulting engineer) wrote: 'To design for inversion (conversion) was in fact to design for failure; it was bad practice.' He said of such an approach to design: 'That way lay danger . . .'.

Durability

In Section A.3.5 of the Annex on Resistance to Chemical Attack, there still remains a sweeping and unspecific claim to the effect that: 'Good quality calcium aluminate (HAC) cement mortars and concretes made in accordance with this annex are resistant to chemical attack'. This theme is elaborated in Section 6.3, which claims 'a much greater resistance than Portland cement concrete against numerous aggressive substances...' and lists some of these, including seawater. In some cases, the claim is valid, although Portland cement concrete performs well in seawater. What is not said, however, is that HAC is always vulnerable to alkalis carried by percolating water. In the revision, there is a proposed a useful addition of a warning that, 'due to the risk of migrating alkalis, Portland cement should not be repaired with HAC in environments with relative humidity above 80%'.

Regrettably, the Annex contains no warning about the danger of corrosion of steel reinforcement in HAC concrete. This has been described in a number of recent papers from the Building Research Establishment (BRE); for example: 'the potential for corrosion in future is probably higher [in HAC concrete] than in other forms of concrete construction' [5]. This finding is not new: in 1963, Rüsch wrote: 'Concrete made with high-alumina cement does not offer the steel as good a corrosion protection as concrete made with Portland cement, as the alkalinity of the former is quite low' [6].

In Section A.6.2, 'Concreting in hot weather', the enthusiastic statement that 'concreting in hot weather can be carried out with no risk...' has been modified by removing the words 'with no risk'. Likewise, the claim that by curing with cold water, it is possible to 'concrete at temperatures around 40 °C' has been replaced by the recommendation to use an admixture 'to ensure long enough working time when concreting in hot temperature'. This is not wrong, but such a statement is not informative and may induce a sense of false security.

Final strength

A note to Section A.7.1 about a rapid method of determining the 'minimum strength' after conversion has been deleted, but the principle of storing concrete at 38 °C for 5 days for the purpose of estimating the long-term minimum strength is still present in BS EN 14647. I doubt the validity of this principle. Section A.7.1 does not tell us whether the concrete should be stored wet or dry, and yet Collins and Gutt [7] reported that wet concrete at 20 °C, or perhaps even occasionally wet, may reach a strength of 10 to 15 MPa lower than when stored dry. It is not clear that there is no moisture effect at 38 °C. Moreover, the same authors [7] report that the loss of strength continues even in highly converted concrete, and even at $w/c = 0.25$, and certainly at 0.4. This is crucial to any determination of the 'final residual properties' required by the Building Regulations [4], and compliance with Building Regulations is at the heart of using HAC in structural concrete.

Influence of w/c on strength

It has been known for almost a century that an increase in w/c results in a decrease in the strength of concrete made with Portland cement. The same applies when HAC is used. There is a difference, however, in that after conversion the decrease is greater than before HAC has undergone conversion. I have discussed this earlier in *Properties of Concrete* [8].

Admittedly, the Annex stipulates a maximum w/c of 0.40, but it is a fact of life in construction on site (as distinct from laboratory made specimens) that occasionally the actual w/c is higher than specified. In consequence, it is important to know the magnitude of the effect of too high a w/c. The revised Annex provides such information by way of a graph plotting the compressive strength of test cubes against w/c. The data in the graph come from two sources; a 1990 paper by George (the manufacturer's chemist) and my book, *Properties of Concrete*, Fourth Edition, published in 1995 [8].

Actually, the table in *Properties of Concrete* is a metricated version of my paper to the Institution of Civil Engineers [3], published in 1963, and at that time severely criticized by Lafarge Cement. My test results are shown in Fig. 3.2.1. However, all's well that ends well, and I am happy that the revised version of BS EN 14647:2003 acknowledges the significant loss in strength with an increase in the w/c in concrete with converted HAC.

Interestingly, there is a plot of the influence of w/c on strength of concrete after conversion in *Properties of Concrete* that uses George's original data so that both Lafarge Cements and I rely on the Lafarge test results. I am pleased that we now sing from the same hymn book, and also that some of my work on HAC is recognized by way of being an additional reference in the revised BS EN 14647:2003.

There is a minor point abut the definition of the w/c: BS EN 14647: 2003 uses the *total* w/c whereas in the last 20 years or so it has become customary to use the *effective* w/c, that is, to ignore the water on the surface of aggregate, which is additional to the saturated-and-surface-dry condition. The latter approach is preferable as the surface water varies from case to case.

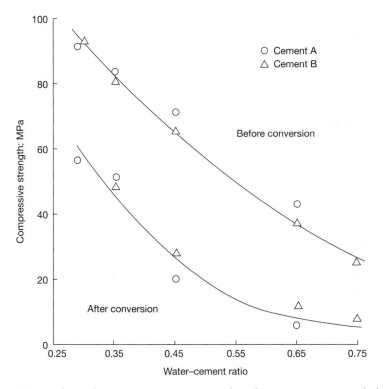

Fig. 3.2.1. Relation between compressive strength and water–cement ratio before and after conversion [3]

The strict requirement for structural concrete with a maximum value of w/c of 0.40 is very helpful, but it is not 'historical' as claimed in the 'Introduction' to the Annex. Even in 1961, the cement manufacturer's handbook stated that a 1:2:4 mix 'normally requires a w/c of 0.5 to 0.6'. The need for much lower w/c was recognized only after the failures in the 1970s.

I know that hindsight improves knowledge but the earlier advice should not be denied. Crossthwaite wrote in paragraph 191 of [3]: 'Engineers in this country had...been under the impression...that, provided special precautions were taken in pouring the concrete... loss of strength would not take place.' And further: '...it should now be clear to all that loss in strength may almost always be anticipated.'

Ten years before the failures of the 1970s, Crossthwaite wrote in paragraph 192 of [3]: 'It was disturbing to note that manufacturers still pressed the sale of HAC without drawing proper attention to the risks attendant on its use.'

Conclusions

Overall, we should welcome the very considerable improvement in the revised BS EN 14647. Nevertheless, the inclusion of an annex, albeit called informatative, about the use of HAC in structures is unusual for a cement standard, as well as inappropriate. This is particularly so in the case of the UK where HAC is not within the ambit of structural design codes. In this country, a mechanism for the use of HAC in structures is provided by the informative Building Regulations via Approved Document by the Secretary of State, 1999, in which HAC is mentioned as an example [4]. What is required by the Approved Document is 'proof of likely satisfactory performance under a given set of circumstances' [4]. The proof is based on 'final residual properties, including their structural properties, ... estimated at the time of their incorporation into the work'.

It is worth noting the use by the Approved Document of the term 'final residual properties' and not just strength, and also the need for the properties during the 'expected life of the building'; thus, carbonation should be considered [4]. Whereas the revised BS EN 14647 is helpful in estimating residual strength, advice on the loss of durability in consequence of conversion is scanty. Admittedly, a figure has been added to show the very much higher porosity after conversion (double at w/c of 0.5 and nearly double at 0.4 as compared with the pre-conversion situation) but the implications of this are not elaborated or even stated explicitly.

On the positive side, we should welcome the unequivocal recognition of the fact that conversion is inevitable under any conditions of exposure. However, structural use of HAC should be considered only in the light of the Building Regulations.

Overall then, I feel that the discussion of BS EN 14647 may well have helped to produce a better Standard. Input from a broad body of technical people is helpful to a standard written by anonymous committee members. The importance of producing the best possible standards cannot be overestimated.

References

1. Neville, A., Draft Standard for high-alumina cement, *Concrete*, July/August, 2003, pp. 44–45.
2. British Standards Institution. BS EN 14647:2003. *Calcium Aluminate Cement – Composition, Specifications and Conformity Criteria*, London.
3. Neville, A.M., A study of deterioration of structural concrete made with high-alumina cement, *Proc. Instn. Civ. Engrs*, 25, July, 1963, pp. 287–324, and Discussion, 28, May, 1964, pp. 57–84.

4. Approved Document to support Regulation 7, HMSO, 1999.
5. Crammond, N.J. and Currie, R.J., Survey of condition of pre-cast high-alumina cement concrete components in internal locations in 14 existing buildings, *Magazine of Concrete Research*, 45, 165, 1993, pp. 275–279.
6. Rüsch, H., letter to the author, 1963.
7. Collins, R.J. and Gutt, W., Research on long-term properties of high alumina cement concrete, *Magazine of Concrete Research*, 40, 145, 1988, pp. 195–208.
8. Neville, A.M., *Properties of Concrete*, Longman, London, 1995.

3.3 SHOULD HIGH-ALUMINA CEMENT BE RE-INTRODUCED INTO DESIGN CODES?

Abstract

A review of the background to the deletion of high-alumina cement (HAC) from British structural codes in 1975 is followed by consideration of the present state of pre-1975 structures, including their strength and durability. The attempts at re-introducing HAC into structural codes are discussed and an argument against such a change is presented. It is suggested that recent changes in Building Regulations have resulted in a satisfactory solution in that designers may use HAC but only when they have demonstrated at the outset of construction that changes in properties of the materials with time will not detract from the performance of the building.

Introduction

'Publications more than 25 years old are likely to be forgotten, which is a shameful waste', was a remark made in *Nature*, on 15 August 1991. However, this lapse of time since high-alumina cement (HAC) was removed from CP110 in 1975 is not the reason for the present paper. It is the recent attempts to re-introduce HAC onto the structural scene that require consideration by structural engineers, especially as the promoters of HAC are not members of our profession.

To answer the question posed in the title requires reviewing the background to the deletion of HAC from structural codes, as well as changes since that time and, of course, looking at the arguments for change. The present state of structures, including strength and durability, needs to be considered. Finally, the impact of the recent changes in Building Regulations will be assessed.

Brief background

In 1973–1974, there occurred three collapses of prestressed concrete roof beams containing HAC, fortunately without loss of life. Investigations that followed established the causes of the collapses. A review of these failures and of the problems with other buildings containing HAC concrete – some 50 000 of them – was published in a book by Neville [1].

The underlying factor in the collapses was the deterioration of HAC concrete, consequential on what is known as conversion of hydrated cement. In essence, conversion is a change in the crystal structure of calcium aluminate hydrates (from hexagonal to cubic) which entails a substantial increase in porosity. As the void content in concrete exerts a direct influence on its strength (an increase in voids of 1% causing a decrease in strength of about 5.5%) conversion leads to a substantial decrease in strength of concrete. The situation is complicated because the rate of conversion influences the degree of loss of strength, and also because at a later stage some of the voids induced by conversion become partially filled by hydration of the hitherto unhydrated part of the cement grains. The presence of water is essential to the progress of conversion, but the interior of concrete is usually wet, and humid conditions may exist in service. Higher temperature greatly increases the progress of conversion.

The increased porosity of converted hydrated cement facilitates the ingress of deleterious agents and, thus, adversely affects the durability of HAC concrete. Now, HAC is much more resistant to acids than Portland cement but, after conversion, it becomes liable to attack known as alkaline hydrolysis. The attacking alkalis may come from overlying Portland cement or plaster, if water is allowed to percolate through these and into HAC concrete. This way, conversion has doubly undesirable consequences.

This section is not concerned with chemical or physical changes due to conversion but there is one aspect that is relevant and yet not mentioned when HAC is advocated. This is bond of converted calcium aluminate crystals. The published information is confusing.

A 1975 paper by George (the manufacturers' chemist) says: '... there is evidence that they (converted crystals) are stronger bonding materials than the hexagonal hydrate (unconverted crystals) since, when neat cement pastes of equal porosity are compared, those hydrated in the cubic converted form show greater mechanical strength than those hydrated in the metastable form' [2]. He cites as basis for this statement a paper on neat HAC paste published in France in 1970.

On the other hand, a 1971 paper by Mehta and Lesnikoff [3] says that the specific surface of hexagonal crystals is about 10 times that of converted crystals, and it follows that this is responsible for part of the loss of strength on conversion. They acknowledge the role of the voids around the cubic crystals, but also point out their smooth surface [3].

A 2000 book by Odler [4] also says that the intrinsic bond properties of converted crystals are inferior to those of hexagonal crystals. He continues to say 'that even at equal porosity a hardened paste containing C_3AH_6 and AH_3 (converted crystals) has a lower strength

65

than one consisting of hexagonal calcium and aluminate crystals' [4]. This is in direct contradiction to the views of George. It seems, then, that George's claim of superiority of bond of converted HAC is not supported by independent researchers.

Causes of failures

Reinforced concrete is a patient material in that it can frequently tolerate one shortcoming but, when more than one untoward factor is at play, failure may occur. As we do not know what will go wrong, do not let us deliberately introduce anything problematic, and HAC is problematic.

In the three collapses of the early 1970s, detailed studies established the relevant factors.

In the failure of the roof over the swimming pool at Sir John Cass's Secondary School, Stepney, Bate of the Building Research Establishment (BRE) who led the investigation, wrote about 'a combination of two factors' [5]: 'Firstly, the HAC concrete had converted considerably with a substantial loss of strength and, as a consequence, its resistance to chemical attack was reduced; it was then attacked by sulphate which led to its disruption' [5]. From a later (1984) perspective, Bate wrote that the failure of the two prestressed concrete beams in Stepney 'was soon identified as being caused primarily by loss of strength of the HAC concrete due to conversion aggravated by chemical attack' [6].

The swimming pool roof in the Stepney school had been exposed to hot and humid air but, importantly, Bate went on to say that, in the adjacent gymnasium roof that had not been exposed to abnormal temperatures, the loss of strength was also unexpected. On the same topic, the Department of the Environment (DoE) wrote: 'It is evident that the problem is not, as was originally thought possible, confined to buildings with abnormally high roof temperatures and humidity' [7].

The letter from the DoE had attached to it an interim statement from the Building Research Establishment (BRE), which says that the fact that in the gymnasium roof (which was not wet and hot) 'the margin of safety provided is insufficient to accommodate any other possible adverse effect, e.g. chemical attack as in the swimming pool or the effects of fire, means that construction using isolated members should be regarded as suspect' [7].

Establishing whether a beam should be treated as isolated is a matter for a structural engineer. In this connection, Bate's closing words are relevant: 'The test results . . . provide some guidance to the condition of structural members of HAC concrete but that engineering judgement must be applied to their interpretation' [5]. The present paper pleads for engineering judgement about the use of HAC in structures.

In the case of the failure of the roof of the Assembly Hall at the Camden School for Girls, the Department of Education and Science [8] concluded that the cause of the complete collapse was a combination of five factors. Three of these were: insufficient bearing of the roof beams; termination of the prestressing wires within the span of the roof beams; and reduction in strength of the concrete in the prestressed roof beams resulting from the conversion of HAC [8]. These words are quoted because of some recent attempts to blame the failure on inadequate bearings to the exclusion of loss of strength. For example, Cather [9] says 'that the failure in one of these collapses had principally been caused by inadequate supports'. With more than one factor at play, we should ask what triggered the failure. It is suggested here that it was the strength of the concrete that changed with time and this is why failure was delayed, so the loss of strength is arguably the culprit.

The influence of the inadequacy of bearings on the shear strength of beams is discussed later in this section.

In the third collapse – the roof of the Bennett Building of Leicester University – I was involved in the physical investigation on behalf of the consulting engineers acting for the contractor. It is, however, the original designer's view that is of paramount importance. He wrote: 'The main factor, therefore, causing collapse was the inadequate tensile strength of the concrete in the edge beams. Investigations have shown that the strength of HAC concrete in the edge beams had reduced to a minimum crushing strength of 6.9 MPa as against a specified strength of 55.2 MPa' [10]. He acknowledges the lack of reinforcement in the concrete under the seatings, but continues to say: 'Had the tensile strength not fallen to well below the estimated value, there is no reason to think that failure would have occurred' [10].

The triggering effect of the loss of strength of HAC concrete was confirmed in a letter to me from the consulting engineer acting for the University of Leicester: 'in a building constructed with the same detail but of Portland cement concrete, distress at the seating was also noticed, although not to the point of failure as in the beam at Leicester' [11].

It may be relevant that the Leicester specification said that prestressed concrete beams should be 'kept damp' for 24 hours, but 'on no account are the units to be allowed to become wet after the initial 24 hour curing period...'. This requirement is unrealistic in a roof.

One more collapse in England, although not spectacular, may be worth mentioning because it predates the other collapses: this is the Sheephouses Reservoir, where the roof consisting of prestressed concrete beams collapsed in 1962. The chemical analysis for the structural engineer led to the conclusion: 'Failure of the beams is most probably due to loss of strength consequent on the change in the nature of the calcium aluminate hydrates on prolonged storage under humid conditions. This process is accelerated by high temperatures' [12]. Interestingly, the cement manufacturers concluded that: 'The failure was caused by alkaline hydrolysis of HAC in the beams...' [13].

Attempts to reintroduce HAC into codes

The preceding section provides the background against which we can judge the possibility of reintroducing HAC into design codes. It is not the objective of this section of the book to argue against a reintroduction as such, but rather to point out the need for a proper justification, provided by a body independent of commercial interests, and not funded by such interests, demonstrating that HAC is safe under *realistic* conditions of construction and service.

The two main moves are a report of a Working Party of the Concrete Society [14] and a draft European standard for HAC [15]. These are complemented by papers emanating from the cement manufacturers and by some press articles.

The Working Party says that it 'concludes, as others have, that a safe basis for the use of CAC (high-alumina cement), including structures, is a reasonable and desirable aim, to take advantage of the beneficial properties of the cement in applications where other materials may have disadvantages'. We should note the word 'structures' and yet the Working Party did not include a single member of the Institution of Structural Engineers. This is a serious lacuna, even though the chairman of the Working Party states that structural engineering advice was obtained from office colleagues [16].

The second move to re-introduce HAC into structural use is in the draft British Standard BS EN 14647, headed Calcium Aluminate Cement (a new name for HAC). The standard is concerned with cement, but it contains a nine-page Annex, entitled 'Essential principles for use of calcium aluminate cement in construction works' [15]. The

Annex discusses the protection of reinforcement, curing and concreting in cold and hot weather, and mentions large beams, 350 by 1800 mm in cross-section. The bibliography contains no reference to any past structural or durability problems, and the references are limited to three works by chemists and one paper by an engineer, dated 1975. The Annex is no more appropriate in a standard on cement than CP 8110 would be as an annex to the standard on Portland cement.

An indirect approach to promoting the return of HAC to structural use is to review and minimize past problems and to say that the three collapses were isolated mishaps, and that everything that followed was an over-reaction. Was it? For example, the minutes of an Institution of Structural Engineers Special Study Group, dated 9 July 1975, show that of 44 buildings inspected, 20 had been strengthened, 4 had shown chemical attack, 1 had safeguarding strength added and 24 had been cleared subject to monitoring or inspection. The numbers speak for themselves; it is also appropriate to note that it was estimated by BRE that there were about 50 000 buildings containing HAC. There were collapses in other countries, and even a fatality in Spain as recently as November 1990 [17].

There is also a ridicule of the existing exclusion of HAC from design codes; for example, in 1977, *New Civil Engineer* described this exclusion as 'a typically British "virtual ban" on HAC' [18], and did not mention exclusion in the rest of the world except France. *New Civil Engineer* then went on to say 'And there are still thousands of HAC structural elements giving good service to this day' [18]. This is true, but many buildings have been strengthened and continue to be monitored. To say that *most* structures are in a good state is not a sound basis for engineering design; and yet *New Civil Engineer* says that 'converted HAC concretes can be durable and provide adequate protection to reinforcing steel – provided the mix contains at least 400 kg/m^3 of cement and has a w/c [water–cement ratio] no higher than 0.4' [18]. This view is not fully supported by a French study [19] discussed later in this section.

Why discuss past failures?

The recent attempts to resuscitate structural use of HAC include a re-interpretation of causes of past failures. Such attempts occurred in the past, too. For example, in the *Civil Engineer*, 21 February 1974, there was a headline on the Stepney failure saying 'The old bogey of conversion can probably be ruled out'. Of course, structural investigation did not confirm that view.

After the British collapses, George [2] wrote: 'There is no justifica-tion for fears about the unpredictability of HAC concrete based on unfathomed dangers...' and 'any inherent dangers associated with its misuse can be avoided by respecting the w/c ratio limit'; this limit is given as 0.4 (and, I would emphasize, not 0.40). What is worrying is that the title of the paper is 'The structural use of HAC concrete' [2] and the author is the cement manufacturers' chemist.

In 1999, Cather wrote: 'an approach exists which can form the basis of the safe use of the material (HAC) in structures' [9]. This is a position much more advanced than that expressed by him in 1997 (with which I agree) when he wrote: 'we are at present still developing procedures for predicting values for long-term *minimum* strength *in situ*' [20] (italics are in the original publication). This is unexceptional. Cather is also right to say that HAC is 'quite unlikely to return to main-stream construction role' [20]. If this is accepted, there is no need to clamour for the return of HAC to the design codes; this section argues later that reliance on a recent amendment to the Building Regulations provides a sensible avenue for the use of HAC when proven to be satisfactory.

Unfortunately, Cather is not right in saying, in 1997: 'The possibility of conversion and provision to account for it were included in structural codes from the 1930s' [20]. Specifically, the *Explanatory Handbook on the Code of Practice for Reinforced Concrete* by W.L. Scott and W.H. Glanville, 1934, contains no warning about conversion of HAC at room temperatures, but only a warning that at 35 °C the strength 'may be only 25% of normal temperatures'. There is a diagram of strength development of HAC concrete with time, and no loss of strength is indicated.

As recently as 2001, Scrivener ascribed the failures in the UK and in Spain to 'major design faults' and the 'external sulfate attack' coupled with 'the use of an excessive water–cement ratio' [21]. This is at variance with the original reports by the structural engineer [6]: unless Scrivener has new and direct evidence, Bate's conclusion should not be altered. Re-interpretation should be based on a proper scientific basis including *structural* assessment of the situation. Just to give one's opinion is unsound at best and may be thought to be biased.

The re-writing of the forensic studies of the 1970s failures continues. In 2000, Dunster and Holton wrote: 'Investigations of the failures showed that, although strength loss had occurred, the primary causes of collapse were poor construction details and chemical attack of the concrete' [22]. No reference to the investigations is given. It could, of course, be their desk study but opinions about 'construction details' have to be linked to structural action. In my opinion, it is for structural engineers, and not materials scientists alone, to opine on the structural use of HAC. This view is supported by Bate who, in his 1984 paper on assessment of structures with HAC, repeatedly referred to appraisal by engineers [6].

It would be a pity if structural engineers abdicated their role in the appraisal of structures. Understanding structural action is crucial; for example, Bate [6] refers to the freedom from failures due to conversion alone being 'the result of and strengthening effects of interaction with adjacent construction'. This is a province of the structural engineer.

What is worrying is that a frequent repetition of the 'new' explanations, even in the absence of any references or documented evidence, might become accepted by those who never read the original evidence about the failures. This could be dangerous.

What has changed since 1975?

A powerful argument for a change in the status of HAC would be a change in its properties. BRE has developed a blend of HAC and ground granulated blast-furnace slag, baptized 'BRECEM' [23]. This obviates the problems of conversion but also removes the high early strength: in broad terms, with BRECEM it takes one year to reach the same strength as HAC alone achieves at one day. This is probably the reason why BRECEM has not been commercialized.

While use of HAC alone in urgent repairs is undoubtedly advantageous, there is evidence from various airports that replacement of parts of a runway can be effected using Portland cement and a superplasticizer, with the slabs back in service within eight hours. So, cost is likely to be the governing factor, and the unit cost of HAC is three to four times higher than that of Portland cement.

Even earlier, Povindar K. Mehta [24] filed a patent for a method of stopping or significantly reducing conversion of HAC, even under hot-wet conditions, by adding calcium sulfo-aluminate ($4CaO.3Al_2O_3.SO_3$) to HAC. This method does not seem to have been exploited commercially.

Thus, conversion and its consequences are still inherent in HAC. The consequences include vulnerability to chemical attack and corrosion of steel. Now, the main use of HAC in the past in the UK was in prestressed concrete, and corrosion of prestressing wire is more serious than corrosion of reinforcing steel. The reason for this is that

71

the diameter of the wire is smaller than the diameter of bars, so that the disappearance of, say, 1 mm thickness represents a greater proportion of cross-section of the steel than in large reinforcing bars.

What has been happening in other countries? It appears that no country has revoked or introduced HAC into structural codes. In 1991, the Strategic Highway Research Program of the American National Research Council produced a major state-of-the-art report on high-performance concretes. The report stated tersely: 'HAC based concretes are not considered because of the conversion of the hydrated aluminate compounds that occur over time, resulting in dramatic increases in porosity and permeability and significant strength loss' [25]. That remains the American view.

The state of existing structures

In considering possible future structural use of HAC, it is useful to examine the health of the numerous structures, built prior to 1975.

BRE provides assessment of such structures as do various engineers. For the protection of clients, reports are rarely published or are not exhaustive. For example, Dunster and Holton say: 'Visual inspection of elements is a reliable method of monitoring general condition...' [22]. I have inspected numerous structures of various kinds containing HAC, including hotels and department stores, and I am convinced that the appearance of a structural element is not a sufficient basis for saying that it is not seriously deteriorated. More importantly, access to many floor and roof beams is exceedingly difficult and, quite often, impossible. Furthermore, I found that sometimes access becomes impeded by new work, e.g. the installation of a sprinkler system or of a false ceiling, which the owner does not want to be disturbed.

One item of advice given by Dunster and Holton is theoretically sensible but unrealistic in practice: 'The building owner was advised to ensure that the areas remained dry' [22]. I have always given similar advice, and repeatedly warned owners of the dangers of wetting HAC concrete, even occasionally. What is seen on the next inspection? A bath had overflowed and water had descended and wetted four floors, travelling sideways here and there. In another case, the roof had leaked but, because there was snow on it, nothing was done for several weeks. So, alas, wetting is a fact of life. People who are not intimately involved in construction and in actual structures in service may not appreciate the extent and variety of problems. In comparison, life in a laboratory is orderly and predictable.

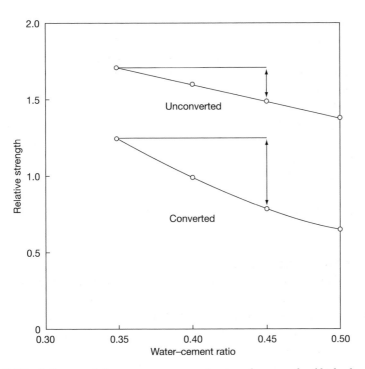

Fig. 3.3.1. Influence of the water–cement ratio upon the strength of high-alumina cement concrete, before and after conversion, relative to the strength after conversion of concrete with a water–cement ratio of 0.4 (based on [26])

The periodic inspections deal with concrete at least 30 years old, and it is arguable that by following the current advice of cement manufacturers to use a maximum w/c of 0.4 and minimum cement content of $400\,\text{kg/m}^3$, all will be well. However, the maximum w/c should be 0.40, and not 0.4 because, to an engineer, 0.4 implies a tolerance of ± 0.05.

Some support to this view is given by Bate's statement that in the Stepney school 'the w/c may have exceeded the maximum of 0.4 aimed at in production' [5]. This is possible, and it is a fact of life that here and there on site, or now and again, w/c may exceed the maximum aimed at. Of course, the same can happen when Portland cement is used but the consequences with HAC are much more severe than with Portland cement. Figure 3.3.1 shows the loss of strength of HAC concrete on conversion; this figure was plotted from the original data in a paper by the cement manufacturers [26]. If we consider the reduction in strength between w/c of 0.35 and 0.45 for

HAC, this can be as large as 17 MPa [27]. Now, for Portland cement, the reduction is typically 10 MPa. So the consequences of too high a value of w/c are more serious in HAC. In engineering language, we can say that there is a hazard of too low a strength *both* in Portland cement concrete and in HAC concrete, but the risk is higher with HAC.

There is a further important point about the w/c. Once the concrete has hardened, there is no simple and reliable way of determining what is the w/c in a structural element. A 1998 BRE paper [28] reports values of w/c from weighing a dry core in air and a saturated core in water and in air and from the cement content and bound water content, but it gives no information on precision or accuracy, and there is no British or ASTM standard for the determination of w/c of hardened concrete.

Furthermore, Collins and Gutt, also of BRE, reported on HAC concrete that 'it is very difficult to measure with any accuracy the original w/c of concretes taken from structures, although a detailed analysis of one such concrete suggested a free w/c in the range of 0.45–0.55' [29]. It seems, therefore, that we can determine the w/c of hardened concrete only within ±0.05 [30]; the same applies to Portland cement concrete [30].

With respect to strength, laboratory tests at BRE tell us that wet concrete at 20 °C or 'perhaps even occasionally wet HAC' may reach a strength of 10 to 15 MPa lower than when stored dry [29]. Given the experience of unexpected 'floods' in buildings, the warning is worrying. The loss of strength continues even in highly converted concrete, and even at w/c of 0.25, and certainly at 0.4.

In a paper in 1997 in the journal of the Institution of Structural Engineers, R.N. Hill [31], a consulting structural engineer, wrote that 'poorly maintained HAC concrete roofs are generally in a weaker condition than protected internal HAC concrete floors', and he cited two examples of problems. In one of these, a secondary support system was introduced; in the other 'X-beams were observed to be very fragile' and the roof structure was taken down [31]. A shell roof was found to be adequate and Hill said 'that HAC concrete roofs were not always straightforward to assess or predict...' [31]. He concluded by saying that owners may be unaware that they have HAC concrete roofs and expressed the view 'that, as structural engineers, we have a responsibility to act in the public interest... and that it would reflect poorly on us if another HAC concrete roof failed so long after we had first addressed the problem' [31]. These sentiments are well worth bearing in mind.

Indeed, in this litigious age, the professional consequences of failure could be disastrous.

Factors other than compressive strength

In assessing existing structures, the strength of cores is not the only concern with respect to structural behaviour. The mode of failure of beams is also relevant. It is well-established that reinforced concrete beams are designed so that the reinforcement governs, with the consequence that failure is slow and progressive, which provides observable warning to occupants by way of deflection and cracking of soffits or ceilings. Now, HAC concrete generally fails suddenly, as was the case in all past failures. We should note that the BRAC sub-committee P rules [32], which we still think to be good, do not apply to isolated beams but only to floors and roofs with fill-in pots and bonded screed, in which load shedding and sharing takes place. In this manner, an overloaded or distressed beam can be partially relieved of the load on it. This is helped by end and side restraints but these are smaller in roofs than in floors.

There are other aspects of structural behaviour that are influenced by a reduction in strength of HAC. Specifically, creep of concrete is proportional to the stress-strength ratio [27]. As the strength drops, this ratio increases, and so does creep and deflection. There may be an adverse effect on finishes and partitions.

These are the sort of considerations that should be addressed in structural research before HAC concrete can be deemed to be just like Portland cement concrete in design codes. On the other hand, if HAC is treated as a special case, the designer can deal with any problems on a case-by-case basis.

With respect to shear, Currie and Crammond [33] quote earlier tests by Bate, which showed that the shear capacity of HAC concrete beams is affected by the bearing width. On the other hand, studies by BRE reported to the Institution of Structural Engineers Special Study Group in 1976 indicate that, in specific X-shape beams, there was no significant difference in the shear strength between bearings of 50 mm and 100 mm [34].

Currie and Crammond [33] point out that the shear capacity of beams is sensitive to the bar at the centre of the cross-section because they do not normally have shear reinforcement. They say that 'early indications of shear failure may be given by horizontal cracking in the web near the supports...' [33]. Unfortunately, in many inspections, I have found great difficulties in getting access to supports, and we must remember that shear failure is sudden.

Durability

Nowadays, the durability of concrete structures needs to be considered explicitly. The first factor is the corrosion of steel, preceded by carbonation. In 1994, Currie and Crammond said that 'the risk of corrosion is becoming an increasingly important consideration in the assessment of HAC components' [33].

Corrosion is, of course, not limited to HAC but, originally, we were led into thinking that, because of the absence of $Ca(OH)_2$ in hydrated HAC, there was no problem of carbonation and of the consequential corrosion.

The extent of carbonation in existing HAC concrete structures has been established only relatively recently by BRE, and this should be acknowledged here. Carbonation of concrete may lead to corrosion of steel, but the link between carbonation and the risk of corrosion is not straightforward. According to BRE, it does not follow that, if concrete is carbonated, structurally significant corrosion will occur [35].

Interesting data on corrosion were obtained in Spain on 1400 joists made in the period 1940–1950 and examined in 1991 [17]. It was reported that carbonation had been accelerated in concrete containing 'converted (HAC) due to high porosity' [17]. The survey in Spain showed that 15% of these joists were made with HAC, the rest with Portland cement. All the HAC joists were carbonated right through the thickness of cover as against 18% of joists made with Portland cement. Portland cement concrete joists exhibited corrosion and cracking just as HAC joists did, but in different proportions: 20% of HAC had an active corrosion rate (against $12\frac{1}{2}$% of Portland cement) and 9% of HAC showed cracking (against 5% of Portland cement) [17].

It is rational to ask why corrosion of steel did not always lead to cracking. It is likely that the corrosion products diffused into the pores and voids in the concrete and precipitated there. Now, converted HAC has pores that may accommodate these corrosion products; such beneficial effects of porous concrete made with Portland cement have been observed by me. The same explanation was suggested as a possibility by Broomfield *et al.* [17].

Crammond and Currie confirmed that, because of increased porosity after conversion, 'the potential for corrosion in the future is probably higher than in other forms of concrete construction' [36]. They said that in over 90% of internal beams examined, both damp and dry, carbonation reached the prestressing steel, and recommended a determination of the depth of carbonation in structural appraisals [36]. The same survey of 14 buildings of various kinds, between 20 and 35 years old, concluded that there is 'a high possibility that the reinforcement in components will no longer be protected from corrosion' and that there is 'the probability that corrosion may be present in buildings more than 25 years old, even in internal environments' [36].

Despite the above, in 2000, Crammond and Dunster wrote that 'the structural consequences of corrosion are unlikely to pose any serious concerns for public safety, since most buildings and structures would exhibit warning signs of distress before significant loss in structural capacity occurred' [37]. Apart from the evidence, presented earlier in this section, to the effect that corrosion does not always lead to cracking, there may be concern about the soundness of advice to look out for 'signs of distress', and about the reference to 'most buildings'. Also, as mentioned earlier, the evidence may be hidden or the failure sudden.

Another problem that may arise is the attack of converted HAC by alkalis, usually carried by percolating water from material above the HAC concrete. Crammond and Dunster investigated a school in which they found deterioration due to alkaline hydrolysis with carbonation where roof leakage had occurred [37]. What happened was that sodium or potassium salts from Portland cement above HAC interacted with the atmospheric carbon dioxide and attacked the hydrated HAC to form calcium carbonate and hydrated alumina, reducing the cement paste to powder. In addition, sulfates from Portland cement led to deposition of ettringite and gypsum in cracks or on the surface, and this contributed to further disruption of the HAC beams [37].

What is surprising, and described as such by Crammond and Dunster, is that sulfates were *also* found on a beam in what they describe as 'nominally dry locations'. They ascribe this occurrence to 'some minor moisture movement' [37]. This means simply that dry conditions cannot be guaranteed, just because there is no visible water. Sensitivity to water is an inherent weakness of HAC in structures.

Damage due to alkaline hydrolysis is not common, but BRE reported an interesting case [38]: 'Cleaners had regularly disposed of cleaning water into a gulley directly above the beam'. The beam was isolated and suffered complete disintegration, with 'practically no reinforcement left in the lower flange of the beam' [38]. This situation is relevant to the exhortation to keep the concrete dry. The *BRE Digest* [38] says that in this way 'The possibility of chemical attack is almost eliminated...'. But who will oversee the cleaners at 6 a.m. and 9 p.m.? Or detect a leaking roof before water has descended into offices or rooms lower down?

In a bridge in France, where HAC was used in the repair of a jointing element between beams made with Portland cement, the use of de-icing salts led to alkaline hydrolysis, a lowering of pH and deterioration due to the formation of aluminium hydroxide [19]. The HAC was fully converted (when examined at the age of 14 years) and had a porosity of 6.4% by volume. This observation is relevant to the consideration of the use of HAC in bridges and in bridge repairs in situations where de-icing salts may be applied.

I. Weir of Parkman Consulting Engineers confirmed the presence of continuing durability problems in existing buildings containing HAC, but noted that 'only a very low percentage of clients were aware of the ongoing durability issues associated with the potential for further chemical deterioration' [39].

Non-structural uses

HAC has many special applications, the main one being in refractory concrete. The various uses are outside the scope of this section, which is limited to structures.

However, it is useful to mention that in small conduits for drinking water, aluminium can be leached in excess of the amount recommended by the European Union [40]. The Concrete Society does not recommend such use of HAC [14], but Scrivener and Capmas of the cement manufacturers limit themselves to saying the 'careful assessment should be made of the possible impact of aluminate leaching' [41].

Apologia

Why have I felt justified in writing the present section? The early work on conversion of HAC at high temperatures was done at BRE. In 1951, *Digest* No. 27 [42] said: 'HAC should not be used in places where it will be both moist and at a temperature of about 85 °F (30 °C) or higher. These conditions acting together cause a loss in strength, whether they occur early or late in the life of the concrete.' My first paper on the strength loss at warm temperatures was published in 1957 [43], other papers following in 1958 [44], 1959 [45] and 1960 [46]. In 1963, I published a major paper on the general topic of deterioration of structural concrete made with HAC [47].

The 1959 paper [45] ends with the words: 'The tests . . . point out the possibility of dangerous deterioration of structural members.' In 1960 [46], it was reported that, at w/c of 0.29, the strength after conversion was reduced from 93 MPa to 55 MPa, and at a w/c of 0.65, from 44 MPa to 5.3 MPa.

I was a member of the Institution of Structural Engineers Committee that wrote the *Report on the Use of HAC in Structural Engineering* [48] and also of the Institution Special Study Group on structural use of HAC concrete [49]. I was invited by the Department of the Environment to serve on the Building Regulations Advisory Committee, Subcommittee P, on HAC chaired by Bernard Stone, whose report is still the main basis for assessment of floor and roof joists [32]. In fact, the 2002 revision [50] of the Stone Report confirms the original rules of the assessment of existing structures, a stage at which much is known about actual loads and construction details. This situation is quite different from design. For existing buildings to be deemed safe, the proviso of absence of chemical attack should not be forgotten.

In view of the above, as well as of a book [1] on HAC published in 1975, it is arguable that, in major discussions on the *structural* use of HAC, my views should have been considered and, if thought fit, refuted. And yet a 70-page chapter on HAC by Scrivener and Capmas [41], listing 193 references, does not even mention Neville as the man who is wrong; in the terminology of George Orwell, I am an 'unperson'.

Anyway, I am not an implacable foe of HAC. *New Civil Engineer* [51], in 1992, in an article headed 'Neville speaks up for HAC', quotes: 'It is engineers who are responsible for their structures, and not the scientists, who should be allowed to make the decision on whether to use HAC again.'

Codes of Practice were economical with their advice; for example, CP 114:1957 said: 'HAC concretes are sometimes unsatisfactory in warm moist conditions', and referred the user to the manufacturers of cement. The previous edition of CP 114:1948 simply allowed the use of HAC.

In addition to codes, advice on structural use of HAC was proffered by the cement manufacturers, and indeed (as in 1957) some codes directed users to the manufacturers' advice. This varied. For example, in 1963, the Cement Marketing Company [52] recommended limits on the *maximum* cement content, which were a function of the member size, and ranged from 260 to 400 kg/m^3. Nowadays, the manufacturers recommend a *minimum* of 400 kg/m^3.

Concluding remarks

It should be made clear that this section is not advocating a ban on the use of HAC in structures. The current absence of HAC from CP 8110 should remain. A mechanism for the use of HAC is provided by the 1999 Amendment to Building Regulations, Regulation 7 [53]. This allows the use of 'adequate and proper materials which are appropriate to the circumstances in which they are used . . .'. The practical guidance is given in 'Approved Document' by the Secretary of State, 1999 [54]. In a section headed 'Materials susceptible to changes in their properties', in which HAC is mentioned as an example, it is said: 'Such materials can be used in works where these changes do not adversely affect their performance.' The proof is 'that their final residual properties, including their structural properties, can be estimated at the time of their incorporation into the work. It should also be shown that these residual properties will be adequate for the building to perform the function for which it is intended for the expected life of the building' [54].

The onus is then on the designer to take a decision on a case-by-case basis. We can note what one designer wrote in 1997: 'Thus I fear it will continue to be an uncomfortably brave action for an engineer to specify HAC concrete in normal structural application' [55]. We can also recall that the then Secretary of the Standing Committee on Structural Safety wrote in 1995 that the previous advice on the structural use of HAC concrete was an error [56].

Offering a warning about potential hazards is a professional engineer's duty; this was highlighted at a conference of the Royal Academy of Engineering in 1991 [57]. The ethical duty of engineers to issue warnings was confirmed by John Uff QC, FREng [58]: 'The papers contain other examples of warning delivered but acted on too late, including the case of High Alumina Cement (HAC) on which Professor Adam Neville CBE FREng FRSE had given clear published warnings in the 1960s. Despite this, a series of collapses in the 1970s still took the industry and the public by surprise' [58].

Nevertheless, the manufacturers of HAC advertise vigorously their product, including, as in Argentina, the uses in various kinds of civil engineering construction. In the UK, their recommendations are invariably for a minimum cement content of $400 \, \text{kg/m}^3$ and a maximum w/c of 0.4 (but not 0.40).

Now, let us assume for a moment that concrete with these limiting values is satisfactory for structural use. Should it be restored to the ambit of CP 8110? As far as prestressed concrete is concerned, this would be unwise because of the potentially serious consequences of an accidentally high w/c in terms of strength of the structural member and of corrosion of prestressing steel. In reinforced concrete, the situation is less critical but, in practice, the production control of reinforced concrete on site is generally less good than in factory production of prestressed concrete units.

The durability problems discussed in this paper militate against routine use of HAC in structures. The amended Building Regulations permit the use of HAC on the basis of engineering judgement, backed by proof of likely satisfactory performance under a given set of circumstances.

So, things should be left alone. Research on HAC will of course continue. When it has come up with new developments, then there will be the case for a new look. Until then, let sleeping dogs lie.

References

1. Neville, A., with Wainwright, P.J., *High Alumina Cement Concrete*, The Construction Press, Lancaster, 1975.
2. George, C.M., *The Structural Use of High Alumina Cement Concrete*, Lafarge Fondu International, Neuilly-sur-Seine, France, 1975.
3. Mehta, P.K. and Lesnikoff, G., Conversion of $CaO.Al_2O_3.10H_2O$ to $3CaO.Al_2O_3.6H_2O$, *The American Ceramic Society*, 54, 4, 1971.
4. Odler, I., *Special Inorganic Cements*, E & FN Spon, London, 2000, pp. 86–199.
5. Bate, S.C.C., *Report on the failure of roof beams at Sir John Cass's Foundation and Red Coat Church of England Secondary School, Stepney*, Building Research Establishment, CP 58/74, June, 1974.
6. Bate, S.C.C., *High alumina cement concrete in existing building superstructures*, Building Research Establishment Report, 1984.
7. *Interim Statement by the Building Research Establishment: Collapse of roof beams at Stepney*, Department of the Environment, London, 30 May, 1974.
8. *Report on the collapse of the roof of the assembly hall of the Camden School for Girls*, Department of Education and Science, HMSO, London, 1973.
9. Cather, R., Letters to the Editor: High alumina cement, *Concrete International*, 21, 3, 1999, pp. 7–8.
10. Samuely, F.J. and Partners, Preliminary comments on Messrs Ove Arup and Partners report on the investigations into partial roof collapse in the Geography room, University of Leicester, Bennett Building, May, 1974.
11. Hobbs, R.W., Letter from Ove Arup & Partners, 22 November, 1974.
12. Williamson, N.W., *Failure of prestressed concrete beam and hollow block roof at Sheephouses Reservoir*, June, 1962.
13. *Report on the Bacup Reservoir roof collapse*, prepared jointly by APCM Research Department and Lafarge Aluminous Cement Co. Ltd, 1963 and received by Department of the Environment, 1975.

14. *Calcium aluminate cements in construction: a re-assessment*, Technical Report 46 of a Concrete Society Working Party, The Concrete Society, 1997.

15. Draft British Standard, BS EN 14647, *Calcium Aluminate Cement – Composition, Specifications and Conformity Criteria*, 20 March, 2003.

16. Neville, A., *Neville on Concrete*, American Concrete Institute, Farmington Hills, MI, 2003.

17. Broomfield, J.P., Rodriguez, J. *et al.*, Corrosion rate measurement and life prediction for reinforced concrete structures, in *Structural Faults and Repairs Symposium*, June, 1993, pp. 155–163.

18. Parker, D., Case for conversion, *New Civil Engineer*, 23–34, 29 May, 1997.

19. Deloye, F.-X., Lorang, B. *et al.*, Comportement à long terme d'une liaison 'Portland-Fondu', *Bulletin des Laboratoires des Ponts et Chaussées*, 202, March–April, 1996, pp. 51–59.

20. Cather, R., Calcium aluminate cements – a perspective, *Magazine of Concrete Research*, 49, 179, 1997, pp. 79–80.

21. Scrivener, K.L., Historical and present day applications of calcium-aluminate cements, in *Calcium Aluminate Cements*, R.J. Mangabai and F.P. Glasser, Editors, IOM Communications, 2001, pp. 3–23.

22. Dunster, A. and Holton, I., Assessment of ageing high alumina cement concrete, *Structural Survey*, 18, 1, 2000, pp. 6–21.

23. Osborne, G.J., BRECEM: a rapid hardening cement based on high alumina cement, *Proc. Instn Civ. Engrs Structs and Bldgs*, 104, Feb., 1994, pp. 93–100.

24. Mehta, P.K., Preventing loss of strength in concretes made with high-alumina cements, Patent application, USA, 1966.

25. Zia, P., Leming, M.L. and Ahmad, S.H., High-performance concretes, *Strategic Highway Research Program*, NRC, Washington, DC, 1991.

26. George, C.M., Manufacture and performance of aluminous cement: a new perspective, in Mangabhai, R.J., Editor, *Calcium Aluminate Cements*, E & FN Spon, London, 1990, pp. 181–207.

27. Neville, A.M., *Properties of Concrete*, Fourth Edition, Longman, London and John Wiley, New York, 1995.

28. Dunster, A.M., Holton, J.R. and Beadle, A.E., *The performance of ageing calcium aluminate cement concrete: lessons from case studies*, BRE Report 353, 1998.

29. Collins, R.J. and Gutt, W., Research on long-term properties of high alumina cement concrete, *Magazine of Concrete Research*, 40, 145, 1988, pp. 195–208.

30. Neville, A., How closely can we determine the water–cement ratio of hardened concrete?, *Materials and Structures*, 36, June, 2003, pp. 311–318. (Section 2.2 in this book.)

31. Hill, R.N., Are old high alumina cement concrete (HACC) roof structures still a problem?, *The Structural Engineer*, 75, 23 and 24, 1997, pp. 421–422 and 432.

32. Report by sub-committee P (high alumina cement concrete), [the Stone Committee], BRAC Advisory Committee, Department of the Environment, 1975.

33. Currie, R.J. and Crammond, N.J., Assessment of existing high alumina cement construction in the UK, *Proc. Instn Civ. Engrs Structs & Bldg.*, 104, 83–92, Feb., 1994.

34. Currie, R.J., *The ultimate uncracked shear capacity of low strength high alumina cement prestressed 'X' beams*, BRE, Department of the Environment, 1976.

35. Dunster, A. *et al.*, *Durability of ageing high alumina cement (HAC) concrete: a literature review and summary of BRE research findings*, BRE Centre for Concrete Construction, 2000.

36. Crammond, N.J. and Currie, R.J., Survey of condition of pre-cast high-alumina cement concrete components in internal locations in 14 existing buildings, *Magazine of Concrete Research*, 45, 165, 1993, pp. 275–279.

37. Crammond, N.J. and Dunster, A.M., *Avoiding deterioration of cement-based building materials: lessons from case studies:* 1, BRE Laboratory Report, 17 pp., 1997.

38. Assessment of existing high alumina cement concrete construction in the UK, *BRE Digest*, 392, March, 1994.

39. Weir, I., High alumina cement concrete. The long-term legacy – what guidance is needed for owners to manage their buildings?, *BRE Seminar on Condition of HAC Concrete Buildings*, 8 March, 2000.

40. Neville, A.M., Effect of cement paste on drinking water, *Materials and Structures*, 34, 240, July, 2001, pp. 367–372.

41. Scrivener, K.L. and Capmas, A., Calcium aluminate cements, in *Lea's Chemistry of Cement and Concrete*, P.C. Hewlett, Editor, 709–776, Arnold, London, 1998.

42. High Alumina Cement, *Building Research Station Digest*, 27, Feb., 1951.

43. Neville, A.M. and Zekaria, I.E., Effect on concrete strength of drying during fixing electrical resistance strain gauges, *RILEM Bulletin*, 38, Feb., 1957, pp. 95–96.

44. Neville, A.M., The effect of warm storage conditions on the strength of concrete made with high-alumina cement, *Proc. Instn Civ. Engrs*, 10, June, 1958, pp. 185–192.

45. Neville, A.M., Tests on the strength of high alumina cement concrete, *JNZ Engng*, 14, 3, March, 1959, pp. 73–76.

46. Neville, A.M., Further tests on the strength of high-alumina cement concrete under hot wet conditions, *RILEM Int. Symposium, Concrete: Reinforced. Concrete in Hot Climates*, Haifa, July, 1960.

47. Neville, A.M., A study of deterioration of structural concrete made with high-alumina cement, *Proc. Instn Civ. Engrs*, 25, July, 1963, pp. 287–324, and 28, May, 1964, pp. 57–84.
48. *Report on the use of High-alumina Cement in Structural Engineering*, Series No. 39, 17 pp., September, 1964.
49. IStructE Special Study Group, *Guidelines for the appraisal of structural components in high alumina cement concrete*, Oct., 1974.
50. Moss, R. and Dunster, A., *High alumina cement concrete, BRAC rules*: revised 2002, BRE Centre for Concrete Construction, 2002.
51. Neville speaks up for HAC, *New Civil Engineer*, 15 Oct. 1992, pp. 5–6.
52. *Structural high alumina cement concrete*, Note Tec. 1866/R10, The Cement Marketing Co. Ltd, Technical Dept, London, June, 1963.
53. *The Building Regulations (Amendment) Regulations 1999*, Statutory Instrument 1999, No. 77.
54. *Approved Document to Support Regulation 7*, HMSO, 1999.
55. Mathieson, G., Letter to the Editor: HAC uncertainty, *New Civil Engineer*, 12 June, 1997, p. 14.
56. Menzies, J.B., Hazards, risks and structural safety, *The Structural Engineer*, 73, 21, Nov., 1995, pp. 357–363.
57. Preventing disasters, E.C. Hambly, Editor, Royal Academy of Engineering, London, 1991.
58. Uff, J., *Engineering Ethics: Do Engineers Owe Duties to the Public?*, The Royal Academy of Engineering, London, 2002.

DISCUSSION

From the Chairman of the North Thames Branch, Mr Stephen Vary

This paper was presented at a technical meeting of the North Thames Branch of the Institution of Structural Engineers. The contributions were extremely lively and some gems of practical information were put forward. Have you heard of Friday concrete? Read on!

George Mathieson, Consulting Engineer

The possible publication of a European standard implicitly condoning structural use of high-alumina cement concrete is a matter for grave concern. Unfortunately, it seems that the European code committees are subject to a variety of intense political or sectional pressures. Recently a draft code was sent out for comment with a particular criterion for slip resistance. There were, however, three mutually contradictory methods allowed for evaluating this figure. BSI was prevailed on to inform the European standards agency that if this became a Euronom, the British edition would be published with a clear statement in the preface setting out the contradiction in the standard, and stating that only one of the test methods set out was acceptable. Perhaps a similar procedure should be followed in this case, and pressure brought to bear on BSI to make the point that HAC, whether under that name or any other, is not appropriate for normal structural work in the UK.

The question has been raised of the appearance of deteriorating HAC concrete. It can occasionally look thoroughly unsound, though certainly I have more frequently seen obvious deterioration with OPC (Ordinary Portland Cement) concrete. The real problem with HAC concrete is that it generally looks marvellous. It may, however, have only 50% of its design strength. In addition, it is often used in prestressed elements, where the prestressing wires are relatively small and may well have relatively high cover. This corrosion can be completely invisible. One cannot rely on seeing long rusty splits and cracks in the beams as prior warning of collapse. Failure is quite likely to be by shear and preceded by no visible warning.

A significant number of us have the responsibility of going round particular major buildings that have structural members made with HAC. We have to check the strength of the structural members and decide whether, in all circumstances, a reduction in the margin of structural safety of these members, of say 50%, renders them dangerous. The consequence of being unduly conservative would be immense cost and inconvenience to thousands of people. The consequence of being unduly rash could be widespread injury and death. We have quite enough buildings like this, without creating more. I would suggest that all structural engineers should take positive steps to nip this apparent proposal to reintroduce HAC for structural use in the bud. Failure to do so may come back to haunt us.

Dr John Bensted

I am in agreement with the main thrust of Professor Neville's arguments that high-alumina cement should not be banned from all structural work but, if it is used, it should be subject to the conditions given in the 1999 Amendment to the Building Regulations with the final decision being made by the responsible structural engineer. Most engineers and materials scientists are happy with these current Regulations on HAC in structures and would not want to see them watered down in any way.

There are errors in the text of Annex A. Specifically, the idea that calcium aluminate cement has previously been known by the different names – high-alumina cement, aluminous cement and high-alumina melted cement – is incorrect as all three names are still currently in use. Second, the inclusion of an informative annex entitled 'Essential principles of calcium aluminate cement in concrete and mortar' in this cement standard is inappropriate. The terminology 'Essential principles...' implies that the information given is normative (compulsory), whereas in reality it is informative (i.e. advisory). This Annex would be more accurately entitled 'Guide for users...' as was the situation in an earlier draft of the Annex.

The reality is that Annex A should have been produced by the European concrete committee CEN/TC/104 but was, in fact, delegated by them to the cement committee CEN/TC/51, albeit with representation of CEN/TC/104 on CEN/TC/51. Annex A needs to be fully revised so that it is actually of value to users of high-alumina cement in mortar and concrete. It would help if Annex A mentioned where HAC can and cannot be suitably employed. If the comments made are not taken on board by CEN/TC/51 then a UK National Annex might be required to address any outstanding concerns when EN 14647 eventually appears.

Gordon Rose, FIStructE

Professor Neville is right to draw our attention to any possible revival of HAC. 'What would be the point?' The 'benefits' of HAC have long been superseded and the risks of structural failure by using it well known to far outweigh even those 'benefits'.

HAC concrete should be outlawed if possible. It may be that 'Who learns from mistakes of the past is not destined to repeat them' but, despite the optimism of some at the meeting, experience also tells us that once an item gets into a Eurocode or CEN system, it is highly unlikely that it will come out. We should prepare for the worst.

One of the most difficult aspects of concrete made with HAC is the finding and monitoring of it, and we are still paying for that.

As Professor Neville explained, Portland cement did not come from Portland. The name was a piece of Victorian marketing that, by equating the colour of the concrete made using such cement with that of a tried and trusted material, sowed in the minds of the customer and public an association with the proven quality of Portland stone. Today such a device may well be challenged, and all need to know is that, although the colour of HAC concrete may be similar to Portland cement concrete, their properties are vastly different.

That history contains the seed of a suitable fallback position, so that Public Engineers and, most important, Insurers, will know where their risk lies. Any code or standard, European or British, should make it mandatory that HAC producers introduce into their produce a distinctive colour before it is sold in the UK.

Ian MacPherson

I am the 'other person' present from the original Stone Committee that Professor Neville mentioned earlier in his talk. At the time, I was professional head of Building Regulations in the then DOE, so sat in on the deliberations of the Stone Committee; so to speak as the clients' 'man'. It is good to see Professor Neville in such fine form.

I recollect spending an 'away weekend' with Professor Neville and the rest of the Committee at the then PSA training establishment at Cardington. It was a weekend best described as 'interesting' because it emerged (remember, this was some time ago) that I was the only person present who had brought a pocket calculator. It was, therefore, made clear to me that I had better tuck myself away and work out the statistics of a mass of data that had been sent to the Committee. So my weekend was spent extracting square roots the hard way, since this was before pocket calculators were fitted out with a square root key. But out of this came a pretty clear realization that the length of the tail of the distribution of concrete strengths was not exactly good news. I took the figure into the Committee and they, after some thought, made some heroic decisions. In the event (30 years of experience?), their heroic decisions turned out to be correct, because, mercifully, no building has fallen down in this country with loss of life, to the best of my knowledge.

There are two issues that flow from that.

The first is that we have available an unrivalled corpus of information on the way in which semi-probabilistic structural concepts might work in practice, since the range of values is far wider than one would ever try out in a test regime involving real buildings. As far as I know, nobody has followed this up, but it would be very apposite to the present thinking that is going on concerning the calibration of the Structural Eurocodes.

The second issue is the success of the inspection regime smartly instituted as a result of the leading recommendation of the Stone Committee. As far as I know, nobody has followed this up either, as an exploration of the administrative aspects of dealing with structural failure on a large scale. The inspection regime itself raised some interesting aspects. The first of these was that it emerged that a large quantity of high-alumina cement concrete had arrived into buildings without anybody realizing. Precast prestressed floor beams arrived on site, were seen to be grey, and of the right length, and some came with a piece of paper that suggested that they were of the required strength. There was no need to (or if there was, it was not observed) tell the structural engineer or architect that the floors were made of high-alumina cement rather than Portland. In consequence, the only sound information as to where the products had gone lay in the manufacturers' files. The manufacturers were, in the event, extremely helpful. But, somewhere along the line, someone suggested that the repair costs for all this work could be recovered by suing the manufacturers. I did not think this idea had four legs, but, just to check up, sent one of the structural engineers up to Companies House to look at the filed accounts of the various manufacturers. It was interesting to observe that the search was promptly picked up by one of the leading technical journals. It was immediately clear that there was no money to sue for, and that if we did, we would then have to deal with the receivers to obtain the vital technical information on distribution. I chalk up this aspect because it is one that people might like to bear in mind if we ever run into a comparable situation again.

An interesting thought emerged a year or two later, as part of a 'state visit' to what was then perhaps the leading research concrete institute in the world. I can, I think, mention the conversation we had after a good lunch, because the institute itself has long gone, and the people concerned are no longer active. Somewhere around the brandy, I naively suggested that the institute must have known about the problems that high-alumina cement was capable of generating, and wondered why they had not murmured in people's ears about this. The instant reply was that they were only funded to research into Portland cement and that that was all they had ever looked at. 'But', said I, 'I was shown before lunch what must be one of the best technical libraries in this subject anywhere in the world. Presumably, you will have read the papers?' There was a pause, and then someone, looking up from the table, said: 'Lovely day for the time of year, isn't it?'

Rowland Morgan, Research Fellow, Department of Civil Engineering, University of Bristol

Adam Neville has demonstrated how immersed he has been in HAC for such a long time. This is not the first time he has raised professional concerns about HAC at a meeting of an international engineering body. Dr Neville knowing what he did about HAC appealed to, what was for him, the highest technical court in the country, the ICE, with some crucial papers in the late 1950s and early 1960s and yet no one in authority seems to have taken any action – even to be cautious!

The HAC fiasco in the 1970s provided marvellous teaching material for feeding student engineers as, indeed, Adam had done to me and my fellow students in Manchester University 20 years earlier.

Structural engineers need to be clear about their several roles within the construction industry and resolve clearly their engineering relationships to codes and standards and specifications whether they be contractors, designers, suppliers, researchers or teachers or whatever. So this is the politics of engineering resulting from all sorts of conflicts that need to be resolved and, sadly, the spin that often goes with it – like changing names so that HAC becomes 'calcium aluminate cement'. It was rumoured that the British government in the 1950s refused to ban HAC for structural use because it would financially damage only one manufacturer – and a French one at that, which was politically unacceptable.

There is another message for young engineers about too much innovation in one project tempting too many things to go wrong. But now we have listened to Dr Neville we cannot avoid a message of responsibility for what happens next. Certainly we cannot do nothing. I can still remember some 45 years later Adam Neville telling us students about the HAC floored factory having a change of use to sweet manufacturing which caused higher temperatures and humidities than the concrete could withstand. Professionalism and ethics are complicated. We may be tempted to accept that publication of some of the discussion on Dr Neville's lecture in the *Structural Engineer* will satisfy some collective responsibility but how we ensure that as Chartered Engineers we fulfil our individual responsibility is a question I leave for Adam Neville to answer.

Written contribution from Douglas Boyd, Retired Consulting Engineer in Belfast, was read at the meeting

I enjoyed reading Adam Neville's paper in the last 2003 issue of the Journal, and approved the conclusion.

I would expect Adam Neville to be aware of a further deficiency in HAC. If not, I report my experience. In 1945 I was Agent/Engineer for a contract in Lincoln where some hundreds (perhaps 700) of piles were cast on site using HAC and driven by drop hammer. HAC was specified by Mouchel because ground condition was ash fill and sleech over peat to gravel. (The pile caps, up to $180\,\mathrm{ft}^3$ were also to be cast in HAC until I asked for instructions on how to keep them cool.)

The phenomenon was this. Piles, which were 12×12 in. in cross-section, could be driven without problems when up to three days old. From four days to (from memory) ten days old, the heads kept shattering at an alarming rate, maybe as many as 80% of them. After that there was no problem, and throughout the 3–20 day period there was no reduction in cube strength.

Some years later, I met some Lafarge people at an exhibition at Olympia and asked them to explain this behaviour. The only comment was 'I believe that we may have heard of it before: would you like something to drink?'

In later years, as a consulting engineer I have assessed many cases of HAC in floors and beams. On one occasion, Friday afternoon, I had just got the test results for compression on a couple of Intergrid beams in a college workshop. I phoned the principal and, in effect, said don't use the workshop. (We installed a support system.)

Response from Adam Neville

At the meeting, in reply, Adam Neville reviewed the thermal history of the piles as compared with the cubes. The surface/volume ratio of the piles was 3 times smaller, if the cubes were 6 inches in size, and $4\frac{1}{2}$ times smaller if they were 4 inch cubes. Consequently, the heat loss from the piles was much smaller than from the cubes. Moreover, the cubes were cured in water at, probably, 20 °C and could lose heat by conduction, so that the temperature rise in the cubes was much smaller than in the piles, which could lose heat only by convection. In consequence, the temperature in the piles rose considerably and conversion of the HAC was rapid and so was the resulting loss of strength. If the above is correct, it would explain why the piles disintegrated at ages of three to about ten days. Beyond that age, secondary hydration of HAC led to improved strength but, of course, only temporarily so.

When we started inspections for the structural integrity of buildings containing precast prestressed units, we had to establish first whether they were made with HAC. This turned out not to be as simple as would be expected. We might look at three or four units and find they were made with Portland cement and conclude that the building did not contain HAC. However, we later discovered that there were buildings that contained a mixture of units, some made with HAC and some with Portland cement. The explanation lies in what came to be known as 'Friday concrete'.

HAC was used in making the precast prestressed units because it allowed a re-use of high-quality steel forms every 24 hours. Production was usually on five days a week. It occurred to a clever precaster that the forms used on Friday were not needed until Monday. There was thus no need to use high-early-strength HAC, which was more than twice the price of Portland cement. In consequence, Portland cement was used on Fridays. All the units were stored together and moved to site, not necessarily in the order of production. Moreover, they were not installed in a building in the order of arrival on site. In consequence, a building might contain a variable number of Portland cement units. Hence, if the initial inspection showed, say, three Portland cement units, it was no proof that the building need not be inspected for the integrity of HAC units.

One particular aspect of durability should be emphasized: when an alkaline solution percolates into HAC, especially if this is converted, harmful chemical reactions take place. In consequence, care should be taken when using HAC as a repair material. If it is in contact with Portland cement concrete, there is the possibility of an alkaline

91

solution from the Portland cement concrete entering and disrupting the HAC.

I am heartened by the contributions to the discussion, and I cannot help noting that no supporter of the use of HAC in structures wishes to go on record. The offers of a drink as a means of dealing with a HAC problem, reported independently by Messrs Boyd and MacPherson, make me feel deprived: I was given no such 'moist curing'.

Where do we go from here? In other words, what is the answer to Rowland Morgan's question: how do we ensure that, as Chartered Engineers, we fulfil our individual responsibility?

A helpful suggestion is made by George Mathieson: 'pressure (should be) brought to bear on BSI to make the point that HAC, whether by that name or any other, is not appropriate for normal structural work in the UK'. George Mathieson also says: 'all structural engineers should take positive steps to nip this apparent proposal to reintroduce HAC for structural use in the bud. Failure to do so may come back to haunt us.' The same sentiments were expressed by Richard Hill in the *Structural Engineer* in 1997 when he wrote: 'that, as structural engineers, we have a responsibility to act in the public interest... and it would reflect poorly on us if another HAC concrete roof failed so long after we had first addressed the problem.'

WRITTEN DISCUSSION

A.N. Suryavanshi, Koihapur, India

There has been some apprehension among engineers in the use of HAC in structural concrete when the danger of reduction in strength due to conversion was pointed out. The author Professor Neville must be congratulated for pointing out additional drawbacks such as the concrete becoming porous, enhanced corrosion in steel and greater deflection under creep.

In India, HAC is not used much, although there is no ban against its use. The Indian standard IS 456-2000, *Code of Practice for Plain and Reinforced Concrete*, mentions it under:

5.1.2: High Alumina Cement conforming to IS 6452 (or sulphated cement conforming to IS 6090) may be used only under special circumstances with the prior approval of engineer in charge. Special literature may be consulted for guidance regarding the use of these types of cement.

My experience is that engineers do not try materials for which no clear guidelines are given in the code itself. Neither the engineers nor the clients would accept the task of monitoring the structure, nor accept its furbishing. Neither would they accept the higher cost of the structure due to the higher price of HAC. Minimum use of HAC at $400\,kg/m^3$ will be a further deterrent to its use.

I agree with the author's view that it will be difficult to control w/c ratio of 0.40 at site.

The author has cited some examples of accidental or occasional wetting of concrete made from HAC. Could he please explain under these circumstances, when it accidentally or occasionally becomes wet, if the conversion process of the concrete starts? How long is the period of wetting required to cause full conversion? Will the water used in mixing cause conversion.

The author seems of have no objection in the use of HAC for refractory concrete. But in his paper it is also mentioned that at $35\,°C$ the strength may be only 25% of normal temperature. In that case the use of HAC in refractory concrete may not be advisable.

Response by Adam Neville to the letter from Mr A.N. Suryavanshi
I am grateful to Mr Suryavanshi for informing us about the position of high-alumina cement (HAC) in India with respect to structural concrete. It is gratifying that he supports my views. It is interesting to learn that the relevant Indian standard allows the use of HAC 'only under special circumstances with the prior approval of engineer in charge' and that Indian engineers do not use materials for which clear guidelines are not given in codes.

With respect to wetting and the amount of water necessary for conversion to take place, the situation is as follows. Young concrete contains a significant amount of water and, with wet curing, which is strongly recommended with HAC, the conditions can be described as wet. As far as neat cement paste is concerned, Midgley [1] found that at depths in excess of $25\,mm$ the relative humidity was 100% regardless of the ambient humidity. In concrete, it is probably only the outer $50\,mm$ that lose moisture to the ambient medium.

So conversion takes place, at a higher or lower rate, in most concrete elements made with HAC and it is accompanied by a loss of strength. What is worrying is that the loss of strength continues even in highly converted concrete, and even at values of w/c below 0.40, probably even at 0.25. Wetting aggravates the situation: Collins and Gutt [2] reported that wetting results in a strength 10 to 15 MPa lower than when the concrete remains permanently dry.

I am not surprised that Mr Suryavanshi found an apparent contradiction between the adverse effects of temperatures above 20 or 30 °C on the strength of HAC on the one hand, and its use in refractory concrete, on the other. I am sorry I did not explain the situation in my paper, but its topic was the structural use of HAC, and not its general study; this can be found in my 1975 book *High Alumina Cement Concrete* [3] and in *Lea's Chemistry of Cement and Concrete* [4].

The influence of temperature upon the strength of HAC concrete is not monotonic. Between room temperature and about 500 °C, HAC concrete loses strength to a greater extent than Portland cement concrete; then up to 800 °C, the two are comparable; above about 1000 °C, HAC gives very good performance, but the strength falls to well below 25% of the original value. Above 700 to 1000 °C, there is a gain in strength due to the development of ceramic bond between the cement and the aggregate. The behaviour of the cement depends very much on the alumina/lime ratio: specifically, white HAC has a ratio higher than 2.5, and even 4.7 [4]. The cost of such cement is many times higher than the price of even ordinary HAC. Special refractory aggregates, such as fused alumina or carborundum, are used above about 1400 °C. Further discussion would be too specialized for the present purposes and would soon exceed my expertise.

References

1. Midgley, H.G., The mineralogy of set high-alumina cement, *Trans. Brit. Ceramic Soc.*, 66, 4, 1967, pp. 161–187.
2. Collins, R.J. and Gutt, W., Research on long-term properties of high-alumina cement concrete, *Magazine of Concrete Research*, 40, 145, 1988, pp. 195–208.
3. Neville, A. with Wainwright, P.J., *High Alumina Cement Concrete*, The Construction Press, Lancaster, 1975.
4. Scrivener, K.L. and Capmas, A., Calcium aluminate cements, in *Lea's Chemistry of Cement and Concrete*, P.C. Hewlett, Editor, 1998, Arnold, London, pp. 709–776.

4

Durability issues

This chapter comprises five papers on selected durability issues. The scope is perforce limited because the topic of durability of concrete is vast, and concerns about durability have justly overtaken our earlier preoccupation with achieving the requisite compressive strength.

I found it, therefore, necessary to be very selective in choosing the durability issues to be included in this book. My choice was governed by recent developments and preoccupations, and my first-hand experience. In the event, I have limited myself to carbonation, sulfate attack and alkali–silica reaction.

The first of these – carbonation – is discussed in Sections 4.1 and 4.2, and arose directly from litigation. Alas, litigation about concrete structures is a growth industry. While intrinsically such litigation is unproductive and wasteful of engineering and scientific talent, it forces us to focus on topics not previously thought about and therefore not studied. Nevertheless, once a topic thrown up by litigation has entered the public arena, the outcome (if a serious study) is worth sharing with the technical community at large.

Sections 4.1 and 4.2 represent my study of carbonation proceeding from cracks, as compared with carbonation proceeding from an exposed surface of concrete. The findings are quite illuminating.

Sections 4.3 and 4.4 deal with sulfate attack on concrete, a subject of numerous court cases in the USA, especially in California. The remarkable thing is that our knowledge of the phenomena involved and, above all, our means of measurement of the extent of sulfate attack are far from standardized or even understood. While I do not pretend to have solved the problem, at least I can claim to have identified the uncertainties and to have pointed out where research and elucidation are needed. These sections should provide a starting point of a large number of research projects.

The final section, Section 4.5, is concerned with minimizing the alkali–silica reaction in concrete. The approach is well known and has been enshrined in European Standards but, early in 2005, the issues involved have suddenly become topical. The reason for this is bizarre. Cements with specific upper limits of alkalis are marketed in the UK and elsewhere, and this makes the mix design relatively easy. However, without any warning whatsoever, Lafarge – probably the largest cement manufacturer in the world – announced that the alkali content of cements produced in one of its plants (in Wiltshire) systematically reported false values of alkali content. This created a stir in the technical press; for example, *Construction News* 20 January 2005, carried a headline, 'Lafarge workers falsified data on cement supplied to its ready mixed concrete customers: Concrete scare hits south-west'.

Consequently, mix design, implemented with the 'false' cements had to be reviewed. I undertook the task for one large construction project, and this led me to a review of the mix design methods for minimizing the alkali–silica reaction, presented in Section 4.5.

It is worth noting that the title of Section 4.5 uses the expression 'minimizing' the relation, and not preventing it. Designers are familiar with this type of approach, but some engineers and clients seek reassurances that all risks have been eliminated and that the design prevents the occurrence of undesirable phenomena. It is important to educate all concerned that total prevention is either unachievable or would be prohibitively expensive. This approach would not serve us well: engineers are expected to produce an acceptable satisfactory and economic design, and not the best possible solution regardless of cost. The lesson has to be pressed home again and again. The classical example that much of the public does not share the above view is the instant clamour whenever a flood, however unusual and unprecedented, occurs.

4.1 CAN WE DETERMINE THE AGE OF CRACKS BY MEASURING CARBONATION? PART I

'An interesting use of the measurement of depth of carbonation could be in estimating the time since the occurrence of a crack.' These words appeared in the third edition of my book *Properties of Concrete*, published in 1981. They were repeated, without any change, in the fourth edition of that book, published in 1995, and in the subsequent 12 impressions [1].

In the intervening 22 years, no one seems to have picked up the idea; at least, I never came across a paper reporting such a method of determining the age of cracks, or dating them. Not only was there no confirmation of success; there was not even a brief note saying that the method does not work. The only exception is the *Proceedings* of a meeting of the Texas Section of ASCE, held in March 2002 [2]; this is not a widely disseminated publication, and I discovered it only in the course of writing this section.

Then, in 2003, a major investigation was conducted whose objective was to demonstrate that certain cracks had been caused by a recent earthquake. Cores of cracked concrete were taken and the cracks were dated by measurement of the depth of carbonation at right angles to the crack surface. The question was asked: how reliable was the dating? This prompted me to look at all the aspects of the method, with a view to answering the question in the title of this section. This, then, is its genesis.

The discussion of carbonation appears in two parts. Part I (Section 4.1) reviews carbonation and its measurement in general; Part II (Section 4.2) deals specifically with carbonation around cracks and gives my best answer to the question posed in the title.

What is carbonation?

This section is not about the science of carbonation, but understanding the phenomena involved is vital to proper testing and, above all, to the interpretation of observations made during testing.

Whereas chemists are, no doubt, well versed in the details of the reactions starting with gaseous carbon dioxide (CO_2) in the air and leading to the formation of calcium carbonate, engineers are less well informed on the subject, yet it is engineers who are concerned with carbonation. In a nutshell, unlike uncarbonated concrete, concrete that has undergone carbonation almost to the surface of embedded steel no longer protects it from corrosion. This is not the subject of this section, but an understanding of the reactions and of the progress of carbonation is vital to the interpretation of the phenolphthalein tests for the purpose of dating of cracks.

Carbonation is a reaction between CO_2 in the atmosphere with calcium hydroxide, which is a major product of hydration of Portland cement. It is well known that gaseous CO_2 does not react with calcium hydroxide, so the presence of water is essential; at the other extreme, it is believed that carbonation does not progress at the surface of concrete in contact with water containing dissolved CO_2 at a relative humidity of 100%, or nearly that high. Indeed, the relative humidity of ambient air is a major factor influencing carbonation: the rate of carbonation is highest in the range of relative humidity of 50 to 70%.

In 2003, a clear explanation of the phenomena involved was offered by Dow and Glasser [3]. According to them, CO_2 initially dissolves as molecular CO_2 but, depending on the composition and pH of the solution, the CO_2 adjusts to an equilibrium between aqueous CO_2, H_2CO_3, ionized $H_2CO_3^-$, and ionized CO_3^{-2}. As this article is written by a nonchemist for nonchemists, I shall use the umbrella term, mild carbonic acid. Ionized CO_3 and the calcium ions from calcium hydroxide react with one another, and the product, having a very low solubility, precipitates as calcium carbonate, $CaCO_3$. This precipitation takes place at the water–cement interface so that the transport of ions from the cement paste must occur through the calcium carbonate and is, therefore, progressively slowed down. I shall return to this when discussing the rate of carbonation but one can note, in passing, that carbonation densifies the cement paste and, consequently, locally increases its strength [1].

The partial pressure of CO_2 in the air is a factor in the progress of carbonation. Broadly speaking, the concentration of CO_2 in rural air is about 0.03% by volume; in an unventilated laboratory, the content may rise above 0.1%; in cities, it is 0.3% but may rise to 1%. The concentration of CO_2 in the air nowadays is a hot topic but, whatever is done to limit the generation of CO_2 on a global scale, it is generated by people and by motor cars. All of this is relevant to the occurrence of carbonation inside a crack. Also relevant is the observation by Dow and Glasser that, if the gas phase at the air–water interface is stagnant, it may become totally depleted in CO_2 [3].

Measuring carbonation

The simplest test for distinguishing between carbonated and uncarbonated concrete is by spraying phenolphthalein lightly onto a freshly exposed surface. There is no need for a trained petrographer; even I have done it.

Surprisingly, there is no ASTM or European standard for this test. ASTM C 856 refers only to microscopic determination of carbonation; BS 1881:Part 201:1986 limits itself to saying that a loss of alkalinity 'may be detected by the use of a suitable indicator, e.g. phenolphthalein, which can be sprayed on to a freshly exposed cut or broken surface of concrete'. As for specific advice, all that I found is the RILEM Recommendation CPC 18, published in its final form in 1988 [4]. There exists also a British document, Building Research Establishment (BRE) Digest 405, published in 1995 [5]; this is an advisory document. A consequence of the absence of national or international standards is that no single test procedure is generally accepted.

The phenolphthalein test

I have already referred to the absence of standard test methods but, because of this, as much uniformity in procedures as possible is desirable so that test results can be replicated and are comparable.

Not being a chemist, I shall rely on the *New Shorter Oxford Dictionary*, published in 1993, for a description of phenolphthalein: 'a whitish or yellowish crystalline solid, $C_{20}H_{14}O_4$, which is used as an indicator in the pH range 8 (colourless) to 10 (red), and medicinally as a laxative'. The last-named use of phenolphthalein I shall ignore, at least in this book.

In carbonation tests, what we call phenolphthalein is really a solution: according to RILEM CPC 18, one should use a 1% phenolphthalein in 70% ethyl alcohol [4]. However, other strengths of phenolphthalein are also used; for example, *BRE Digest 405* prescribes 1 g of phenolphthalein dissolved in 50 mL of alcohol and diluted to 100 mL with deionized water [5]. What I believe to be important is that the same strength of phenolphthalein should be used in all the tests if reliable conclusions are to be drawn. What is also very important is that no tap water is used because such water would introduce various ions and alter the alkalinity (pH) of the concrete surface, and this is precisely what the phenolphthalein test determines.

The phenolphthalein solution is sprayed lightly onto a freshly exposed surface of concrete; clearly, a surface that has been exposed to the atmosphere for some time would be carbonated, unless the relative humidity of the air is very low or the air is saturated with moisture. There is, however, a question about how the fresh surface should be obtained. There are two possibilities: a sawn and, therefore, smooth surface, or a split and therefore rough surface; this issue is considered in the next section.

This may be an appropriate place to say that the phenolphthalein test does not really measure the carbonation of the cement paste but only its pH. In essence, the application of phenolphthalein to a concrete surface makes it appear pink if the pH is higher than about 9. Somewhere between 9 and 8.2, depending on the circumstances, there is no pink colouration, which indicates carbonated concrete. Between these values of pH, the differentiation is not easy. *BRE Digest 405* describes the situation in clear terms: 'The main disadvantage of the pH indicator spraying methods is that they can differentiate only between fully carbonated concrete and other areas of concrete which might vary between being completely unaltered and being almost fully carbonated' [5].

More importantly, a small amount of alkalis will allow the pink colour to persist. This is significant because, according to Hime, the interpretation of the situation as 'uncarbonated' may be made erroneously, even when 90% of the cement paste at the surface has been carbonated [6]. This is why, when assessing the carbonation front in relation to the risk of corrosion of steel, it is assumed that corrosion may take place ahead of the observed carbonation front [1, 7].

We can note that Lo and Lee observed that 'the carbon dioxide could react at depths much greater than those indicated by the phenolphthalein' [8]. They also said: 'Past estimations of carbonation depth based on phenolphthalein indicator may have underestimated the level of carbonation' [8]. For this reason, 'converting' an observed depth of carbonation into the time since the beginning of exposure to CO_2 can underestimate that time.

Because of the heterogeneity of concrete, the carbonation front is not a straight line; accordingly, RILEM CPC 18 states that, at a given location, generally a graphical average and the maximum depth of carbonation should be reported [4]. The problems of establishing the actual position of the carbonation front are discussed by Houst and Wittmann who say: 'Measurements with pH indicators allow us to define a line which is generally difficult to situate precisely within the carbonation front' [9]. For measurements in an entire structure, Roy *et al.* found that the ratio of the maximum depth to the average depth was 2.5 [10].

Preparation of test surface

I have already quoted BS 1881: Part 201:1986, which, although it gives no detailed procedures, refers to the use of 'cut or broken' surfaces. RILEM CPC 18 states: 'Concrete prisms...that can be split into lengths of roughly 50 mm (2 inches) are suitable' [4]. It also says that 'a slice is broken off for each test'. And further on: 'Measure the depth of carbonation on the freshly broken surface. Saw-cuts are not always suitable.' The RILEM recommendation separately considers specimens from existing structures, and recommends 'using drilled cores taken from completed structures and subsequently split', and refers also to 'concrete surface broken by chisel...' [4].

Now, *BRE Digest 405* says: 'A convenient method is to break off pieces of concrete to give surfaces roughly perpendicular to the exposed face. Sawn surfaces are *not* suitable' [5] (italics are used in the original text). The Concrete Society requires freshly broken surfaces, and points out that 'cut or drilled surfaces may give misleading results because they can expose and reactivate unhydrated cement particles in otherwise fully carbonated concrete' [7].

The European Committee for Standardization (CEN) proposed, in 1997, a method for the measurement of depth of carbonation, but the results were found to exhibit an excessive variability [11], and a standard test method has not yet been published. It is worth noting, however, that the trial tests involved a '50 mm slice broken from prism' and sprayed with phenolphthalein [11].

The documents discussed previously, none of which are mandatory, give a range of advice on the type of surface to be tested: from either sawn or broken, to broken. Such a situation is unhelpful and I should explain why I think that the method of preparation matters. A broken surface, that is, one prepared by splitting using a cold chisel, exposes the concrete in the interior in its actual condition. On the other hand, sawing requires the use of water to cool the saw and to remove the fine material generated by sawing. Moreover, following the sawing it may be necessary to use more water to clean the new surface.

These operations of wet sawing and washing move the fine material from one place to another; in addition, if there are cracks in the sawn surface, the wash water may move some fine material into the cracks. The fine material is largely hydrated cement paste, which contains calcium hydroxide and possibly other alkaline material; this is also the view of the Concrete Society [7]. Furthermore, wetting the concrete may leach some calcium hydroxide or salts of sodium or potassium, which are highly soluble, to the surface. In consequence, a carbonated area may acquire some alkalinity: subsequent application of phenolphthalein would show a pink colour and, therefore, a presumably uncarbonated area. Hime observed spurious alkalinity on the surface of aggregate particles caused by the migration of alkaline material [6].

This spurious observation of uncarbonated concrete would be of particular significance when the purpose of testing is to determine the distance that the carbonation front has reached from a crack and also the extent of carbonation of the crack surfaces themselves. My opinion is, therefore, that, at least for the purpose of dating cracks, only broken surfaces should be used.

For the sake of completeness, I should add that, when breaking a concrete sample is not practicable, samples of concrete dust can be obtained by drilling to successively greater depths; the samples are then subjected to the phenolphthalein test. It is important to be scrupulous in keeping the samples separate because if alkaline material from uncarbonated concrete contaminates a sample, the entire sample will turn pink, giving the impression of absence of carbonation [1].

Another aspect of testing that requires standardization is the time interval between the exposure of a fresh surface and the application of phenolphthalein. As exposure of concrete to air allows carbonation to take place, the application of phenolphthalein should take place soon and at a fixed time interval from the exposure of a fresh surface. For site tests, RILEM CPC 18 says that testing should be done immediately and, if this is not possible, the specimens should be stored in CO_2-free containers [4]. *BRE Digest 405* recommends instant testing [5], and so does the Concrete Society, which recommends a 30-s delay [7]. The latter is my view, too. I am emphasizing this point because I have heard about re-application of phenolphthalein, and this vitiates a comparison between tests at different locations.

Progress of carbonation

As this section is about a possible specific use of the determination of carbonation, and not about the phenomenon at large, I shall limit myself to saying that, under steady conditions, the depth of carbonation increases in proportion to the square root of time of exposure. An expression commonly used is:

$$D = Kt^{0.5} \tag{1}$$

where D = depth of carbonation; t = time of exposure; and K = carbonation coefficient.

Usually, D is given in millimetres (or inches), and t in years, so that K is expressed in millimetres (or inches) per year$^{0.5}$.

It is important to note the proviso of steady conditions. Otherwise, the square-root rule does not apply. For example, when the humidity is variable, with periodic wetting, the rate of carbonation is reduced because CO_2 has to diffuse through saturated pores. Conversely, surfaces protected from rain undergo carbonation at a faster rate. A typical picture of the influence of the relative humidity on the progress of carbonation was given by Wierig [12]. He reported that for concrete with a water–cement ratio (w/c) of 0.6, at the age of 16 years, the average values of the depth of carbonation were: at relative humidity of 100%, 0; at relative humidity of 95%, 4 mm; and at relative humidity of 60%, 15 mm [12]. Microclimate variations within a building should not be ignored [10].

When it comes to dating cracks in concrete by establishing the distance of the carbonation front from the crack surface, the limitation about steady conditions must be borne in mind. By steady conditions I mean, first of all, the ambient relative humidity. There are, however, other factors that also may vitiate a comparison.

I should add that comparisons can be made between two or more cracks in similar members under similar circumstances. In such a situation, if the age of one crack is known, and the distance of the carbonation front at the same depth in both cracks is known, then the age of the second crack can be determined from Eq. (1). There is, however, a second limitation in that the widths of the cracks should not be dissimilar. I shall discuss this factor in Section 4.2. That section will also deal with other aspects of cracks that are relevant to carbonation.

References

1. Neville, A.M., *Properties of Concrete*, Longman and John Wiley, London and New York, 1995.
2. Patty, T.S. and Jackson, D.J., Use of carbonation to date cracks in concrete, *Proceedings*, Texas Section ASCE, Arlington, TX, 27–30 March, 2002, 3 pp.
3. Dow, C. and Glasser, F.P., Calcium carbonate efflorescence on Portland cement and building materials, *Cement and Concrete Research*, 33, 2003, pp. 147–154.

4. RILEM, *Measurement of Hardened Concrete Carbonation Depth*, CPC 18, 1988.
5. Building Research Establishment, Carbonation of Concrete and its Effect on Durability, *Digest 405*, May, 1995.
6. Hime, W.G., private communication.
7. Concrete Society and Institute of Corrosion, Measuring the depth of carbonation, Current Practice Sheet No. 131, prepared by J. Shaw, *Concrete*, Jan., 2003, p. 40.
8. Lo, Y. and Lee, H.M., Curing effects on carbonation of concrete using a phenolphthalein indicator and Fourier–transform infrared spectroscopy, *Building and Environment*, 37, 5, May, 2002, pp. 507–514.
9. Houst, Y.F. and Wittmann, F.H., Depth profiles of carbonates formed during natural carbonation, *Cement and Concrete Research*, 32, 2002, pp. 1923–1930.
10. Roy, S.K., Northwood, D.O. and Poh, K.B., Effect of plastering on the carbonation of a 19-year-old reinforced concrete building, *Construction and Building Materials*, 10, 4, 1996, pp. 267–272.
11. Jones, M.R. *et al.*, A study of the CEN test method for measurement of the carbonation depth of hardened concrete, *Materials and Structures*, 33, Mar., 2000, pp. 135–142.
12. Wierig, H.-J., *Longtime studies on the carbonation of concrete under normal outdoor exposure*, Instut für Baustoffkunde and Materialprüfung der Universität Hannover, Germany, 1984, pp. 239–249.

4.2 CAN WE DETERMINE THE AGE OF CRACKS BY MEASURING CARBONATION? PART II

Section 4.1 reviewed the topic of carbonation of concrete and its measurement in general. This was the background to the question of dating cracks in concrete by establishing the distance of the carbonation front from the crack surface. This section discusses the various factors influencing the progress of carbonation from the crack surface and leads to the answer to the question posed in the title of this section and Section 4.1.

The references at the end of Section 4.2 are numbered consecutively from Section 4.1, as some of the references in Section 4.1 are referred to in this part as well.

Carbonation in cracks

Section 1 ended with a statement about the possibility of comparing the distance of the carbonation front between two cracks in similar concretes under similar circumstances when the age of one crack is known. Sometimes, however, there is no crack of a known age available for comparison purposes, and this leads to the question of whether carbonation near a crack can be compared with carbonation from a formed surface, such as a wall surface, where the age of the wall, that is, of exposure to carbonation, is known with certainty. The question then is: does carbonation progress at the same rate from a crack surface as from a formed surface? The second question is: does the crack width influence the rate of carbonation?

To answer the first question, we should look at the expression for the depth of carbonation, given in Eq. (1), which is repeated here for convenience:

$$D = Kt^{0.5} \tag{1}$$

where D = depth of carbonation; t = time of exposure; and K = carbonation coefficient.

Usually, D is given in millimetres (or inches), and t in years, so that K is expressed in millimetres (or inches) per year$^{1/2}$.

Now, K depends on numerous factors, some intrinsic to the particular concrete, others depending on extrinsic factors. As this is not a review paper, I shall mention the various factors only very briefly. They are discussed in, among others, *Properties of Concrete* [1]; for a fuller list, I am grateful to Hime [6].

105

The fundamental intrinsic factor controlling the rate of carbonation is the diffusivity of the hydrated cement paste, which is a function of the pore system. Now, the relevant characteristics of the pore system are the size distribution of pores and their connectivity. The pores provide space for the mild carbonic acid, discussed previously; the connectivity influences the ease with which the carbonic acid can penetrate the concrete. The structure of the hydrated cement paste depends on the type of cement and on the mixture used, especially the water–cement ratio (w/c).

Another factor specific to the given concrete is its curing history because this influences the extent of hydration and therefore the pore structure.

In comparisons of carbonation at different locations, it is possible that there are differences in curing history, as well as local differences in w/c. It is arguable that these effects can be overcome by taking a large number of measurements.

It is also useful to point out that the texture of concrete at the exposed surface influences the rate of carbonation: specifically, the texture can be more or less dense, or closed, depending on the type of formwork used, or on the method of finishing in the case of unformed surfaces. The influence of texture is particularly relevant in a comparison of the carbonation front from an external surface and from a crack surface; in the latter case, the surface is naturally more open, but it also contains aggregate particles, which are not permeable.

As for extrinsic factors, I have already mentioned the relative humidity at the surface of the concrete, be it external surface or crack surface, and the concentration of CO_2; the latter may be important when considering carbonation from a crack surface. Coverings, such as mortar, tiles, or wallpaper, which may exist on some surfaces, impede, or even prevent, carbonation from the given face. Roy *et al.* found that there was no carbonation when plaster thickness was 30 mm [10]. According to Engelfried and Tölle, carbonation is prevented when the CO_2 diffusion resistance of a covering is at least that of 50 m of air [13].

Influence of crack width on carbonation

The preceding section makes it clear that great care is required in a comparison of the distance of the carbonation front from an external surface and from a crack surface. As cracks differ in their width, it is important to consider whether the crack width affects the progress of carbonation. To put it a different way, if two cracks of the same age differ in width, will the distance of the carbonation front be the same? I believe this question to be of importance because I have seen an assertion that the crack width has no influence on carbonation and, at that stage, explicit evidence one way or another was not available. This is why I propose to discuss this issue in some detail.

Much depends on the hygrometric conditions in the crack. If it is filled with water, the rate of diffusion of CO_2 present in the air to the crack surface is very low: about four orders of magnitude lower than in dry air. If the crack is dry, CO_2 travels inside the crack in gaseous form and becomes dissolved in water at the surface of the crack, thus becoming a mild carbonic acid.

I would expect the supply of CO_2 in the confined space inside a crack to be more limited than the supply to a wall surface. Whereas in a room, CO_2-laden air circulates and thus supplies CO_2 to the exposed concrete surfaces, in a crack there is no force driving the gas. Extending this supposition, in a narrow crack the supply would be smaller than in a wider crack. This argument has to do with the continuing supply of CO_2, so that it does not become depleted, and not to do with the size of the relevant molecules, considered by some people.

Now for experimental evidence, which is scanty: Liang *et al.* [14] state that 'the carbonation velocity is proportional to the square of the crack width'. I presume that 'velocity' means rate, and the rate of carbonation determines the depth of carbonation after a given time of exposure to CO_2. They say further: 'Both the crack width and the permeability in the crack space must be considered.' Liang *et al.* present a diagrammatic representation of the carbonation front over a depth of a crack of constant width; this shows that the distance of the carbonation front from a crack surface decreases with an increase in the crack depth. This is a fairly conclusive statement, but they also say: 'The influencing parameters of the carbonation depth of cracked concrete are very complicated' [14]. Schiessl states that the diffusion of CO_2 through a crack into the interior of concrete depends on the crack width and the permeability of the crack space [15]. The latter depends mainly on deposits inside the crack.

Liang *et al.* also say: 'The diffusion of CO_2 through a crack into the interior of a concrete member depends mainly on deposits' [14]. And further: 'Deposits may originate from the environment (dirt) as well as from the interior of the concrete member itself' [14]. In my opinion, the presence of deposits is of importance in dating cracks and I shall return to the consideration of deposits in cracks later on.

Further support for my view that the crack width influences the rate of carbonation from a crack surface is given by De Schutter who concludes: 'The main parameters [in the extent of carbonation] are the crack width and the crack depth' [16]. I should add that De Schutter's tests were performed on mortar (not on concrete) and that the cracks were preformed by inserts of constant width; thus, the crack surface was more like that of a formed surface and not like the rough surface of a 'natural' crack.

Influence of crack width on permeability

I have not found any further published direct evidence of the influence of crack width on carbonation, but there exist some papers describing studies on the influence of crack width on water permeability of cracked concrete under low pressure. I believe that the permeability of concrete to water, and to CO_2 gas, and also to CO_2 dissolved in moisture, is similar. If this is correct, then factors influencing one of these would affect the other in a similar manner qualitatively, although not quantitatively.

The basis for my opinion is the argument that, theoretically, the intrinsic permeability coefficient of a given concrete should be the same regardless of whether a gas or a liquid is used. However, tests have shown that gases yield a higher value of the coefficient of permeability than liquids because of the phenomenon of gas slippage. This means that, at the surface of a crack, the gas travels along the surface, whereas a thin layer of liquid is stationary. In consequence, in a given concrete, although the greater the gas flow the greater the water flow, the gas travels faster than the water. This is the reason for my saying that the influence of various factors is qualitatively similar. A corollary is that just as water moves through a wider crack faster than through a narrow crack, so does CO_2, but the difference in flow is greater in the case of water.

If the previous argument is valid, then the findings in the paper by Cordes and Bick, dealing with 'hydraulic flow through cracks' and 'also the slow flow moving capillary transport, the latter mostly observed at very fine cracks', are relevant [17].

They report as factors influencing the mechanisms of transport, the crack width and the roughness of crack surfaces [17]. The roughness is relevant to some tests using preformed cracks, which have smooth surfaces, unlike 'natural' cracks.

Natural cracks were induced in tests reported in two papers by Aldea *et al.* [18, 19]. This was achieved by testing cylinders in a manner used in the splitting tension test, but stopping the increase in load when the desired crack width was reached. The following conclusions are relevant to the subject matter of the present section. First, the water permeability of cracked concretes was the same for concretes with a strength of 36 and 69 MPa. Second, the water permeability increased significantly with an increase in crack width. Specifically, for crack widths smaller than 0.2 mm, permeability increased by a factor of nearly 10, as compared with uncracked concrete. For cracks wider than 0.2 mm, the water permeability increased very significantly [18].

In their second paper, the same authors concluded that cracks 'can act as major pathways for water or aggressive chemical ions to penetrate in concrete, enabling its deterioration' [19]. Ionized CO_2 can be an 'aggressive ion'. Further evidence of the influence of crack width is offered by Ramm and Biscoping who reported that 'the flow-through quantities [of water] ... depended clearly on the crack width and the pH value' [20].

Finally, although not presenting original experimental findings, a book by St John *et al.* says: 'The extent of reaction between the environment and the hardened paste in these wide cracks [between 0.10 and 0.15 mm] will depend on the crack width, the severity of that environment and the quality of concrete' [21].

From the previous discussion, I conclude that the crack width significantly affects the progress of carbonation from the crack walls; in consequence, a comparison of the extent of carbonation between cracks of different widths cannot be used to estimate their ages.

Is dating of cracks possible?

To answer this question, we should start with Eq. (1). This relation does indeed make dating possible but there are severe limitations. The most important of these is that the value of K must be the same for the two situations being compared. This means that the comparator must be of known age and its conditions throughout must have been the same as those for the crack whose age is being determined. The same conditions means: the same concrete mix, the same curing, the same surface texture, as well as some of the other conditions listed as influencing the carbonation rate in the section headed 'Progress of carbonation' in Section 4.1. Under such circumstances, a comparison can be made, but it must be between two formed surfaces or between two cracks of the same width.

A comparison cannot be made between a formed surface and a crack surface, if only because the texture is different in the two cases and also because the availability of CO_2 is not the same. My view is supported by Iyoda and Uomoto who wrote that their 'results indicate that the speed of carbonation caused by cracks is less than the speed of carbonation proceeding from the open surface' [22]. This is relevant to FEMA 306, which says: 'If an estimate of the carbonation rate can be made, then studies of the pattern of cementitious matrix carbonation adjacent to a crack can be used to estimate the age of the crack' [23]. This is a big 'if', which is what I have referred to in the preceding two paragraphs.

This does not mean that some dating of cracks is not possible. For example, Patty and Jackson were able to establish in three cases that the cracks were old [2]. This was done mainly by observation of the presence of debris in the crack; such debris would not be present in a crack suspected of having opened just a few days or a few weeks before the observation of the cracks. A new crack would not have amassed a large amount of debris. The same view is expressed by FEMA 306, which says: 'Secondary deposits within a crack such as mortar, paint, epoxy, or spackling compound indicate that the crack formed before the installation of the material contained within it' [23]. As pointed out by Schiessl, an additional impeding material, although not exactly debris, is the products of autogenous healing, which takes place in narrow cracks when wet; this is discussed in a recent book by Neville [24].

FEMA 306 also points out that 'the edges of the material on either side of the crack become rounded over time due to normal weathering' [23] so that a crack with rounded edges is an old crack.

A converse conclusion cannot be drawn. By this I mean that an absence of debris may be due to the crack being new or to the fact that no debris entered the crack. Anyway, this kind of dating does not rely on carbonation. Actually, Patty and Jackson determined the carbonation front from the crack and, in each case, they found a significantly large value; from this, they concluded that the crack must have been old because extensive carbonation does not occur in a period of days or even a few weeks [2]. But, if very little carbonation has taken place, then this can be due to one of two reasons: either the crack is new or the conditions of exposure were such that carbonation could not proceed. For example, the crack was full of water; alternatively, the crack was exposed to a very low relative humidity. Under either of those circumstances, there would be very little or no carbonation.

Conclusions

On the basis of carbonation measurements and of direct observation, it is often possible to say that a crack is old, but it is not possible to say that it is new. When specific information on cracks that can be used as comparators is available, dating of cracks by the carbonation front is possible. However, comparisons between cracks and exposed formed surfaces are not valid.

It follows that, if there is a dispute about whether a crack was caused by a recent event, the party claiming that it was cannot prove its point by carbonation measurements, but the party claiming that the crack antecedes the recent event can generally prove its case.

I started Section 4.1 by quoting from the Fourth and Final Edition of *Properties of Concrete* [1]. Following the present study, were I to write a post-final edition of my book (which I am not planning to do), I would modify my words to say: Under some circumstances, measurement of the carbonation front from crack surfaces could be used to show that the crack is old; when cracks of known age exist, comparisons could indicate the age of the given crack.

References (continued from Section 4.1, page 104)

13. Engelfried, R. and Tölle, A., Effect of the moisture and sulphur dioxide content of the air upon the carbonation of concrete, *Betonwerk + Fertigteil*, Heft 11, 1985, pp. 722–730.
14. Liang, M.-T., Qu, W.-J. and Liao, Y.-S., A study of carbonation in concrete structures at existing cracks, *Journal of the Chinese Institute of Engineers*, 223, 2, 2000, pp. 143–153.
15. Schiessl, P., *Corrosion of steel in concrete*, Report of the Technical Committee 60-CSC RILEM, Chapman and Hall, London, 1988.
16. De Schutter, G., Quantification of the influence of cracks in concrete structures on carbonation and chloride penetration, *Magazine of Concrete Research*, 51, 6, 1999, pp. 427–435.
17. Cordes, H. and Bick, D., Zum Flüssigkeitstransport and Trennrissen im Stahlbeton, *Beton und Stahlbeton*, 86, 8, 1991, pp. 181–186.
18. Aldea, C.-M., Shah, S.P. and Karr, A., Effect of cracking on water and chloride permeability of concrete, *Journal of Materials in Civil Engineering*, Aug., 1999, pp. 181–187.
19. Aldea, C.-M., Shah, S.P. and Karr, A., Permeability of cracked concrete, *Materials and Structures*, 32, June, 1999, pp. 370–376.
20. Ramm, W. and Biscoping, M., Autogenous healing and reinforcement corrosion of water-penetrated separation cracks in reinforced concrete, *Nuclear Engineering and Design*, 179, 1998, pp. 191–200.

21. St John, D.A., Poole, A.W. and Sims, I., *Concrete Petrography*, Arnold, London, 1998.
22. Iyoda, T. and Uomoto, T., Effect of existence of cracks in concrete on depth of carbonation, *Institute of Industrial Science Section 5*, University of Tokyo, No. 319, 1998, pp. 37–39.
23. Federal Emergency Management Agency 306 (FEMA), Evaluation of earthquake damaged concrete and masonry wall buildings, *Basic Procedures Manual*, 1998, pp. 50–53.
24. Neville, A., *Neville on Concrete*, American Concrete Institute, Farmington Hills, MI, 2003.

4.3 THE CONFUSED WORLD OF SULFATE ATTACK ON CONCRETE

Abstract
External sulfate attack is not completely understood. Part I identifies the issues involved, pointing out disagreements, and distinguishes between the mere occurrence of chemical reactions of sulfates with hydrated cement paste and damage or deterioration of concrete; only the latter are taken to represent sulfate attack. Furthermore, sulfate attack is defined as deleterious action involving sulfate ions; if the reaction is physical, then it is physical sulfate attack that takes place. The discussion of the two forms of sulfate attack leads to a recommendation for distinct nomenclature. Sulfate attack on concrete structures in service is not widespread, and the amount of laboratory based research seems to be disproportionately large. The mechanisms of attack by different sulfates – sodium, calcium, and magnesium – are discussed, including the issue of topochemical and through-solution reactions. The specific aspects of the action of magnesium sulfate are discussed and the differences between laboratory conditions and field exposure are pointed out.

Part II discusses the progress of sulfate attack and its manifestations. This is followed by a discussion of making sulfate-resisting concrete. One of the measures is to use Type V cement, and this topic is extensively discussed; likewise the influence of w/c on sulfate resistance is considered. The two parameters are not independent of one another. Moreover, the cation in the sulfate salt has a strong bearing on the efficiency of the Type V cement. Recent interpretations of Bureau of Reclamation tests, both long-term and accelerated, are evaluated and it appears that they need re-working.

Part III reviews the standards and guides for the classification of the severity of exposure of structures to sulfates, and points out the lack of calibration of the various classes of exposure. A particular problem is the classification of soils because much depends on the extraction ratio of sulfate in the soil: there is a need for a standardized approach. Taking soil samples is discussed with particular reference to interpreting highly variable contents of sulfates. The consequences of disturbed drainage of the soil adjacent to foundations and of excessive irrigation, coupled with the use of fertilizer, are described. Whether or not concrete has undergone sulfate attack can be established by determining the change in the compressive strength since the time of placing the concrete. Rejection of this method and reliance on determining the tensile strength of concrete because of 'layered damage' are erroneous. Scanning electron microscopy should not be the primary, and certainly not the first, method of determining whether sulfate attack has occurred. Mathematical modelling will be of help in the future, but at present cannot provide guidance on sulfate resistance of concrete in structures.

Part IV presents conclusions and an overview of the situation with consideration of future improvements. The Appendix contains the classification of exposure to sulfate given by various codes and guides.

Part I: What is meant by sulfate attack?

Introduction

In the title of this section, I toyed with using the word 'topsy-turvy' instead of 'confused' because 'topsy-turvy' is defined in the *Shorter Oxford Dictionary* as 'utterly confused'. This is how the situation with respect to sulfate attack seems to me, but perhaps topsy-turvy is not a proper adjective for a technical book. It is not that our knowledge of the various phenomena and the actions involved have never been clarified: they are understood, at least at the engineering level, but the disputes in the last ten years or so have resulted in many assertions. These, each taken separately, are not necessarily wrong but, taken together, they result in a thoroughly confused situation.

I suppose that at the outset I should justify my attempt to find some general strands in the issue of sulfate attack of actual structures: I am an engineer and not a scientist, and my view is that the behaviour of structures in the field is the professional concern of an engineer.

I am not alone in emphasizing the need for engineering judgement and, therefore, for an engineer to be prominently involved in studies of sulfate attack; DePuy of the Bureau of Reclamation wrote that, in certain circumstances, 'it becomes increasingly important that the engineers have a good understanding of the chemical reactions, factors governing the rate of reaction, and mechanisms causing deterioration in concrete' [1].

Nor am I alone in pointing out that laboratory studies to date are inadequate to translate into an understanding of field behaviour. DePuy wrote: 'Most of the knowledge of sulfate attack has been developed from laboratory studies involving relatively simple chemical systems of pure materials, yet in actual practice the situation is more complicated...' [1]. And, like me, DePuy feels that, 'Although sulfate attack has been extensively investigated, it is still not completely understood' [1].

Scope of this section

I do not claim that the present section will make the situation crystal clear. I am unable to achieve this, in part because I am a civil engineer, and understanding sulfate attack involves not only civil engineering and construction practices, but also chemistry, soil engineering and field and laboratory testing. It is not easy to get the people from all these disciplines to work together; worse still, there is sometimes a tendency for any one specialist to opine on phenomena outside his or her specialism. Unfortunately, some expert witnesses are tempted to express opinions outside their own field of expertise, fearing that admitting ignorance might detract from their standing, or else they are inveigled into opining in 'alien fields'.

Of course, I, too, must try not to fall into that trap, and yet I am writing on the broad issue of external sulfate attack. So let me reassure readers that I intend to limit myself to identifying the issues involved and to pointing out the disagreements but, of course, this will not prevent me from expressing my views, where I have some confidence in doing so.

'Chemists should be studying chemical attack on concrete' is the title of an article by Hime [2], published in *Concrete International* in April 2003. And yet I – an engineer – am writing on a topic that concerns chemical action on concrete. Why?

First of all, Hime himself says that chemists have not been 'doing the chemistry' of chemical attack [2].

Second, I am not expressing any views on chemical reactions, although I shall need to refer to some of them.

Third, the *real* significance of chemical reactions between sulfates and concrete is in the damage to concrete; mere chemical changes that do not affect the health, performance or durability of concrete are of very limited interest to those concerned with real structures in the field, as distinct from laboratory specimens under artificial conditions. I hasten to add that laboratory research is important and essential to advance our knowledge, but the practical significance of that knowledge needs to be examined in engineering terms.

This brings me to the fourth issue: how widespread is sulfate attack? And also: how significant are the consequences of the reactions of hydrated cement paste with sulfates?

The present section is not concerned with internal sulfate attack or with thaumasite attack, which is a separate form of sulfate attack [3]. Nor does the section specifically deal with sulfate attack on concrete in consequence of oxidation of iron sulfide, such as pyrite, in the soil, which is largely an acid attack because sulfuric acid is formed. This can have very serious consequences, but allowing concrete to come into contact with soil containing pyrite is simply bad practice, and can be avoided or prevented.

Definition of 'attack'

At this juncture, it may be appropriate to define the meaning of the term 'attack'. Usually, I find the *New Shorter Oxford Dictionary*, published in 1993, to be a good starting point. The *Dictionary* says: 'a destructive action by a physical agency, corrosion, eating away, dissolution'. I see the adjective 'destructive' as being crucial; in other words, an action *by itself* does not necessarily represent attack. From an engineer's point of view, what matters is what has happened to the concrete: an action that does not result in deterioration or in a loss of durability is not an attack.

My view is supported by some researchers in the field of thaumasite attack. Bensted says: 'It is important that thaumasite formation and thaumasite sulfate attack are properly differentiated' [4]. This distinction is of vital importance: the mere fact of a given chemical reaction occurring is not evidence of damage and of consequential loss to the owner of the given concrete. Bensted's distinction is equally applicable to sulfate attack that does not result in the formation of thaumasite.

Further support to my views is given by W.H. Harrison, who says: 'when the analysis of concrete reveals a high sulphate content this does not necessarily indicate any deterioration although conversely, loss of strength or visible deterioration accompanied by a high sulphate content would be evidence of sulphate attack' [5].

Also, Skalny *et al.* emphasize that 'the presence of ettringite *per se* is not a sign of sulfate attack' [6]. On the other hand, they refer to efflorescence as 'virtually observable damage', and describe efflorescence as being 'predominantly sodium sulfate; occasionally also sodium chloride, magnesium sulfate, other salts' [6]. They say that efflorescence represents severe *chemical* attack, the word 'chemical' being italicized in their book.

My view that efflorescence *per se* is not a proof of *damage* has been expressed earlier [7]. Mehta [8] also says that efflorescence 'should not cause any damage' except under certain circumstances.

Definition of sulfate attack

I have already expressed the view that attack on concrete occurs only if concrete experiences deterioration or damage. This still leaves open the question of what kind of mechanism needs to be involved for the action to be described as sulfate attack. There are two schools of thought.

One school considers sulfate attack to have taken place if sulfates are involved, regardless of the mechanisms acting. The other school limits the concept of sulfate attack to the consequences of chemical reactions between sulfate ions and hydrated cement paste, so that chemical changes in the paste take place. However, if sulfates interact with cement and cause damage to it, but the action is physical, and a similar action can occur with salts other than sulfates, then the damage is considered to be physical attack or physical sulfate attack.

I am aware of the fact that chemical changes are accompanied by physical changes, and I hesitate to enter into the niceties of science. Why then am I raising the issue? Indeed, it could be argued that nomenclature is of little significance. However, in my opinion, in a large and complex field, it is helpful if everybody uses the same terminology when referring to a given phenomenon.

Furthermore, I believe that there is a difference between chemical and physical attack in that the chemical attack involves *necessarily* the sulfate ion. On the other hand, physical attack involves crystallization of salts, of which one example is sulfate, but it is a topic of wider interest with respect to concrete and also rocks. In consequence, in this paper, the term sulfate attack will be limited to chemical attack, but I shall also consider briefly the physical sulfate attack.

I am far from being alone in holding the above-mentioned views. For example, The Concrete Society Technical Report on diagnosis of deterioration of concrete structures [9] recognizes 'physical salt weathering' as a phenomenon separate from sulfate attack, regardless of whether the salts are sulfates or not.

I would also like to quote Collepardi who, in a state-of-the-art review of delayed ettringite attack on concrete, considers sulfate both of internal and external origin but explicitly excludes 'damage which is specifically attributed to the physical sulfate attack such as crystallization of water-soluble sulfate salts' [10]. Mehta, in a paper sub-titled 'Separating myths from reality' [8], says: 'Surface scaling of concrete due to the physical salt-weathering attack is being confused with the chemical sulfate attack. Salt weathering is purely a physical phenomenon that occurs under certain environmental conditions when any porous solid, such as brick, stone, or poor-quality concrete, is exposed to alkali salt solutions including sulfates (but not necessarily limited to sulfates)' [8]. I am not sure that the qualification 'poor quality' applied to concrete is appropriate; rather, the openness of the surface texture is relevant, and this could exist with some types of finishes of any concrete. Mehta reports that 'the physical attack associated with salt crystallization was not limited to alkali–sulfate solutions; for example, it occurs with alkali carbonate solution...' [8].

I have to record that opposite views are also held; for example, Skalny et al. say that 'to draw a distinction between "physical" and "chemical" processes does not serve a useful purpose and will only serve to further confuse engineers' [11]. Being an engineer, I find this solicitude touching, but I do not believe that we, engineers, are already confused and have to be saved from being further confused!

I would like to emphasize that the issue of chemical or physical attack is not just semantic because the consequences of physical attack manifest themselves in a different way from chemical attack. Moreover, prevention of the two types of attack is likely to be different.

Physical sulfate attack

A prevalent form of physical attack is the reversible change of anhydrous sodium sulfate (thenardite) into decahydrate (mirabilite). If crystallization takes place in the pores at or near the surface of concrete, large pressure may develop, with consequent deleterious action. Hime and Mather say that 'precipitation from supersaturated solution may be required' [12]. The term 'may' shows that the phenomena involved are not yet fully understood.

The problem of physical salt attack has been studied for a very long time. In 1929, Lafuma [13] in France reported his experiments on the transformation of sodium sulfate decahydrate into anhydrous form of that salt, and pointed out the significance of the temperature of 33 °C, above which the anhydrous salt is formed in a solution with a low apparent volume. While the minutiae of Lafuma's papers may, or may not, be correct – and they are not within my field of expertise – I have studied them because they recognize clearly that the actions involved are physical and that the chemical action of sulfate ions does not enter the picture.

In 1939, the crystallization of salts in porous materials was studied by Bonnell and Nottage [14] in England. Their tests were performed on a material with a very low tensile strength, saturated with a sodium sulfate solution. They showed that, 'when hydration takes place within the pores, stresses sufficiently high to overcome the tensile strength of normal porous building materials may be developed' [14] and the salt may 'exert a sufficient force to bring about disintegration of the material' [14].

These were experiments involving thenardite and mirabilite (although these names were not mentioned) even though it was not concrete that was tested. The experiments are highly relevant to the consideration of physical sulfate attack on concrete but the paper by Bonnell and Nottage [14] is probably not well known because it was published shortly before the outbreak of World War II.

In 1949, Lea and Nurse [15] studied the crystallization of solids in aqueous systems, using both sodium sulfate and calcium sulfate. They identified a lack of knowledge of the growth of crystals under stress but, with respect to ettringite, they favoured the notion of topochemical reaction.

Consideration of through-solution reactions as against topochemical reactions is a matter for chemists. Lafuma's proposition of topochemical formation is criticized by Brown and Taylor [16] and also by Mehta [17].

The specific behaviour of sodium sulfate salts has continued to be a subject of study. Thenardite has a much higher solubility than mirabilite so that thenardite can produce a supersaturated solution with respect to mirabilite. In 2002, Flatt [18] expressed the view that, at 20 °C, a tensile hoop stress of 10 to 20 MPa develops by crystallization of mirabilite (decahydrate) from a saturated solution of thenardite (anhydrite). Tensile stress of such magnitude would inevitably disrupt concrete, but the pore size influences the ease with which the crystals can grow [18].

According to Flatt, 'the development of crystallization pressure requires supersaturation in the liquid film between the crystals and the pore wall' [18] but he notes that, in concrete, 'unfilled large pores tend to consume supersaturation in harmless growth' [18]. Flatt expresses the view that sodium sulfate can cause more damage than 'just about any other salt' [18].

In a 2002 paper, Brown recognizes the distinction between what he calls 'classic form of sulfate attack associated with ettringite and/or gypsum formation' and 'physical sulfate attack associated with crystallization of sulfate' [19]. He also acknowledges the fact that the involvement of a sulfate is almost incidental to what is essentially a physical attack by saying: 'Physical sulfate attack can be regarded a specific type of salt damage' [19].

The various papers that I have just discussed are concerned with physical actions, the fact that sodium sulfate has sulfate ions being incidental, so that they point toward the concept of physical attack.

Overall then, I think it is useful to distinguish attack that involves sulfate ions, that is, sulfate attack, from a situation where damage is caused *solely* by physical forces, which is physical sulfate attack. I venture to recommend such nomenclature.

How widespread is sulfate attack?

In a paper reviewing the progress in the practical field of concrete in the last 40 years [20] (Section 7.1), I commented that, in some cases, the volume of laboratory research is disproportionate to the practical extent of the problem. I cited the alkali–aggregate reaction, but now I wonder whether this is not also the case with sulfate attack.

The opinion that sulfate attack is not a widespread problem in concrete structures is expressed even by those who have published extensively on the topic. Specifically, a book titled *Sulfate Attack on Concrete* [6] says 'that neither external nor internal forms of sulfate attack are major causes of degradation of concrete structures', and also that 'sulfate-related deterioration is globally minute' compared to damage by corrosion of steel or by freezing and thawing. The same book says – and I agree – that owners of particular structures that have suffered sulfate-induced damage are justified in being concerned. On the other hand, the book contains a number of sweeping statements, elevating the mere presence of sulfates to damage [6] supported by global references such as: *Deposition Transcripts 1996–2000*. I am aware of some of these transcripts, and I find in them also diametrically opposed views, opinions and interpretation.

Moreover, despite numerous allegations about damaged foundations in California, not a single collapse has been reported, nor has any house or building been evacuated.

In 1993, Mehta, who has repeatedly reviewed the problems of sulfate attack, expressed the view 'that sulfate attack is seldom the sole phenomenon responsible for concrete deterioration' [21]. Seven years later, he wrote that 'the threat of structural failures due to sulfate attack...seems to be even less of a threat than that caused by alkali–silica reaction' [8].

At the 1998 Seminar on Sulfate Attack Mechanisms, held in Quebec City, Pierce (who was head of the US Bureau of Reclamation) said that he echoed Haynes' words: 'We do not see problems in the field', and ascribed this situation to the fact that 'we have taken the necessary precautions' [22]. The precautions are incorporated in codes of practice. Literature search indicates that, in temperate climates, there have been very few, if any, actual failures.

The view that there are very few reported cases of structures in service damaged by sulfates in the soil or in groundwater is supported by Tobin [23]: in a technical information note dated 22 May 2002, he refers to his 66 years' experience with residential and commercial construction using concrete, and says: 'Surface damages have been observed, not unlike those that occur due to freeze–thaw weather conditions on exposed concrete, but not a single instance of a progressive attack or a complete failure due to sulfate attack in the true meaning of the word' [23]. Those are the words of a structural engineer.

These views contrast to a large extent with the opinions of laboratory scientists who use accelerated test methods. Often, their objective is to study sulfate attack, so that they have little interest in concrete that performs well. I am not blaming them, but those dealing with real concrete in-service should not be swayed by tales of doom.

There are, nevertheless, some parts of the world where sulfates are problematic. In South Australia, especially in the city of Adelaide, there is an endemic problem, known as salt damp. Essentially, the ground contains sulfate-laden salt; this rises through, or on the surface of, the concrete, or other building material such as limestone or rubble, in house foundations during cool, wet winters. When the water evaporates during hot, dry summers, salt is deposited on the surface of the concrete. In consequence, concrete may be damaged. Prevention in existing buildings made of porous materials is generally uneconomic, and usually a sacrificial render is applied, and renewed periodically.

Parts of the Canadian Prairies abound in sulfates and so do some other areas in several countries. In the UK, sulfates are found in some clayey soils and on so-called brownfield sites, that is, locations of former factories involving chemical processes; these are not natural deposits.

Mechanism of sulfate attack

First of all, it is useful to list the principal reactions of sulfates with hydrated cement paste. Those of main interest are as follows.

Sodium sulfate reacts

Sodium sulfate reacts with calcium hydroxide to form calcium sulfate (gypsum). This reaction proceeds to a greater or lesser extent, depending on the conditions. In flowing water, with a constant supply of sodium sulfate and removal of calcium hydroxide, the reaction may eventually continue to completion, that is, leaching of all calcium hydroxide (which, on a volume basis, is a major product of hydration of Portland cement). Otherwise, equilibrium is reached.

Calcium sulfate formed as described above can subsequently react with C_3A, usually via the formation of monosulfoaluminate, to form ettringite.

Sodium sulfate is the predominant salt involved in reactions with concrete in California.

Calcium sulfate reacts

Calcium sulfate reacts with C_3A to form ettringite, and reacts also with sodium and potassium hydroxides. It is to minimize the reaction with C_3A that Type V cement was developed.

Calcium sulfate is the predominant salt involved in reactions with concrete in England.

Magnesium sulfate reacts

Magnesium sulfate reacts with all products of hydration of cement: the important resulting compounds are calcium sulfate and magnesia. Calcium sulfate can proceed to react with C_3A.

The reactions of sulfates are well known so that there is no need to write here the relevant equations in stoichiometric form. However, it is important to recognize that the end products of the various reactions, if they occur so as to damage the concrete, result in different types of damage. In consequence, different preventive measures are required.

Specifically, in the case of sodium sulfate, the use of cement with a low C_3A content (which becomes calcium aluminate hydrate) minimizes the extent of reaction with sodium sulfate and, therefore, of the formation of ettringite. This may lead to expansion but, as pointed out by Taylor, 'seeing ettringite in a concrete does not necessarily mean that ettringite has caused the stress' [24].

As reported by Santhanam *et al.* [25], gypsum is the primary product of sulfate attack at high concentrations of the sulfate ion. When the pH of the pore water in hydrated cement paste falls below about 11.5, ettringite is not stable and decomposes to gypsum.

122

The formation of gypsum and its effects are still far from clear. As recently as 2000, tests on pure C_3S paste (that is, in the absence of C_3A) showed that gypsum formation 'may be a cause leading to expansion and subsequent cracking' [26], which occurred after about one year in a 5% sodium sulfate solution. The words 'may be' are important. The paper concluded that 'it is not clear that gypsum formation follows either the topochemical reaction mechanism or the through-solution mechanism' [26]. The final words 'More research is needed in this area' [26] have a familiar ring.

Indeed, Mather said: 'I have seen no evidence that the formation of gypsum during sulfate attack on cement paste causes expansion' [27]. He favours the notion of a through-solution formation of gypsum [27]. W.H. Harrison expresses a similar view [5].

In tests by Rasheeduzzafar *et al.* [28] on neat cement paste and mortar specimens immersed in a solution of sodium and magnesium sulfates, damage manifested itself primarily by scaling, spalling and softening, rather than by expansion or cracking; there was a large reduction in compressive strength [28]. Because magnesium hydroxide has a very low solubility and its saturated solution has a pH of about 10.5, C-S-H decomposes, liberating lime. This lime reacts with magnesium sulfate and forms further magnesium hydroxide and gypsum. This reaction continues until gypsum is crystallized out [28]. Magnesium hydroxide reacts also with silicate hydrate arising from the decomposition of C-S-H, and this results in the formation of magnesium silicate hydrate, which lacks cohesive properties [28]. These reactions take place even when the C_3A content in the cement is very low [28].

Thus, it can be seen that magnesium sulfate is more aggressive than sodium sulfate due to the lowering of pH of the pore water in the hydrated cement paste. It follows that a laboratory test using magnesium sulfate is not valid for field conditions where pH is not lowered during the sulfate attack [21].

In this connection, I have to admit that my own laboratory tests [29] on sulfate attack on concrete were performed using magnesium sulfate because, when I started the tests in 1965, I did not realize the peculiar consequences of that salt. I also used a very high sulfate content in the 'attacking' solution (50 000 ppm) as against several thousand ppm encountered in the field. The error of my ways continues to be repeated because, as Hime says, the reaction is faster, which suits PhD students, and the fact that it is different from that of calcium sulfate seems to elude many investigators [2].

A further complication should be mentioned. Magnesium hydroxide, called brucite, forms a protective layer on concrete, unless this is damaged mechanically [30]. This explains the good record of concrete in seawater, which has a high magnesium sulfate content.

It might be thought by some readers that the preceding apparent literature review is too academic for a section purporting to deal with sulfate attack in service. My defense is that the seeming lack of clarity allows one-sided, and therefore biased, views on the lines that 'if gypsum is present, sulfate attack has occurred' or 'if no ettringite has been found, there is no sulfate attack'. Moreover, often experience with, say, magnesium sulfate, is directly transferred to a situation involving sodium sulfate. Conversely, laboratory test results on a single type of sulfate are applied to field situations, in which a mixture of sulfates is present.

With reference to exposure conditions, it is worth noting that carbonation of a concrete surface prior to exposure to sulfates 'reduces the build-up of sulfate within concrete' [5].

Part II: Progress of sulfate attack and preventive measures

This part of the four-part Section discusses the progress and consequences of sulfate attack, mainly under field conditions, and also some preventive measures.

Progress of attack

It is obvious that sulfate attack can occur only when sulfates are present in contact with concrete. But how much sulfate needs to be present for the reactions between the sulfates and the hydrated cement paste to result in damage to concrete, that is, to constitute sulfate attack? We should also ask the question whether the concentration of sulfate in the water in contact with concrete affects the extent of the attack; the extent comprises the intensity and the rate. These depend on the solubility of the given sulfate and on the continuity or otherwise of the supply, that is, on whether the water is static or is flowing.

The preceding paragraph could be interpreted to refer first to sulfates in the soil, but then to continue with consideration of sulfates in water. The explanation of this apparent inconsistency is that dry salts do not react with concrete so that it is the concentration of sulfates in water in contact with concrete that is relevant. In other words, sulfates can move into concrete only in solution.

The rate of attack by sulfates is affected by the strength of the sulfate solution, but the effect becomes small beyond about 1% of sodium sulfate and about 0.5% for magnesium sulfate [31]. This effect is of significance in the classification of aggressivity of exposure, given in the various codes, and is discussed later in this section.

Taylor lists as consequences of sulfate attack: strength loss, expansion, cracking and ultimately disintegration [32]. He further points out that, whereas laboratory studies have concentrated on 'expansion and cracking, field experience shows that loss of adhesion and strength are usually more important' [32]. This accords with my criticism of excessive reliance on individual, and often disconnected, laboratory studies coupled with an absence of in-depth (no pun intended) evaluation of field experience. Similarly, Brown and Taylor say: 'Further experimental work should take into account the conditions existing in the field and be related to field observations' [16].

As for changes in concrete resulting from the action of sulfates, Skalny *et al.* list nine chemical changes, and nine physical changes, but say that 'by themselves, neither (sic) of the above chemical, physical, and microstructural changes are necessarily an adequate sign of sulfate attack' [6]. They say further: 'However, in combination, there can be little doubt' [6]. This is a very sweeping statement because much depends on which 'combination' is involved.

References to the action of sulfates in southern California should not be generalized because, as stated by Skalny *et al.*, soils there contain 'high levels of sulfates, often in the form of gypsum' [6]. As I have already pointed out, the action of calcium sulfate and its consequences are different from the action of sodium and magnesium sulfates. Thus unqualified statements, such as 'one of the characteristic features of severe external sulfate attack is formation of gypsum' [6], are too vague and may be misleading to engineers without specialized knowledge of chemistry.

How to make sulfate-resisting concrete

It would be contrary to the tenor of this section if I were to make a categorical statement of the 'one-size-fits-all' type about how to ensure sulfate resistance of concrete. However, I have come across forceful assertions that, for concrete to be resistant to sulfates, all that matters is a low w/c. An experienced engineer knows that, unless concrete is well compacted and dense, a low w/c is useless. DePuy gave as a first requirement for good resistance: 'a high-quality impermeable concrete. This requires good workmanship (workable mixes, good consolidation, a hard finish, and good curing), and the use of a relatively rich mix with a low w/c' [1]. Most engineers would subscribe to this view, but some laboratory scientists might have little experience of these site-related factors.

I shall deal with w/c in a later section but, by way of introduction to the concept of sulfate-resisting concrete, I have to mention also the use of Type V, that is, sulfate-resisting cement. This will be considered below.

At this stage, I would suggest that there is no single answer to the question, often posed to me: which is more important: w/c or type of cement? The main reason for the absence of a unique answer is that the mechanism of sulfate attack depends on the cation in the sulfate: sodium, calcium, or magnesium. Of course, more than one of these may be present in a single location. Alas, in disputes, an expert may be expected to present a single answer.

Parenthetically, I would like to refer to another problem occasionally faced by an expert. He or she is asked whether it is certain that, under some circumstances, damage cannot occur. Alas, lawyers cannot, or do not want to, accept that a certainty of impossibility of the occurrence of an event does not exist in engineering.

Apropos of making a categorical statement that is universally right, Northcote Parkinson, a 20th century English satirical writer, commented that no one is always right, but that in every organization there is one person who is invariably wrong. That person is very useful as a soundboard for a new idea claimed to be a 'winner': if that person thinks so too, then the idea is wrong!

Use of sulfate-resisting cement

This type of cement (known as Type V) has existed in the USA since 1940, although since then, ASTM has introduced some modifications in the standard requirements for its chemical composition.

The essential feature of Type V cement is that it limits the content of C_3A as calculated by the Bogue method. However, as pointed out by Mehta, Type V cement 'addresses only the problem of sulfate expansion associated with the ettringite formation' [21]. I would expect, therefore, Type V cement to be particularly efficacious when calcium sulfate is the attacking medium, although it could be beneficial with respect to prevention of the formation of gypsum owing to the action of sodium sulfate.

Thus, Type V cement is of no avail in the attack of calcium hydroxide and C-S-H, and the subsequent loss of strength [21]. This situation is relevant to tests on sulfate resistance: do we determine expansion or loss of strength? These tests are discussed in Part III of this section.

We should note that, for exposure to modest amounts of sulfates, codes allow the use of Type II cement with pozzolans as an alternative to Type V cement.

Nevertheless, the importance of using Type V cement should not be underestimated. The Bureau of Reclamation *Concrete Manual* stated in 1985 that 'Concrete containing cement with a low content of the vulnerable calcium aluminate is highly resistant to attack by sulfate-laden soils and water' [33]. Even today, the Bureau of Reclamation is reported to consider 'the C_3A content to be the greatest single factor influencing the resistance of Portland cement concrete to sulfate attack, with cement content second' [31]. Some specific values of the C_3A content in cement and of the cement content in concrete are quoted but I suspect that it is the density of concrete and not the cement content as such that is the real factor. The view that cement content has no effect on sulfate resistance was expressed also by Tom Harrison in 1997 [34].

Recent interpretations of long-term tests at the Bureau of Reclamation will be discussed in a later part of this section.

Although dated, the views of Lea [35] are of importance. He states: 'The resistance of Portland cements to attack by sulphate solutions is known to be related in a general way to their calculated C_3A content, but there are many anomalies in this relationship' [35]. One reason is that the rate of cooling the clinker influences the morphology of the aluminate: if it is in glass form (amorphous) it is much less susceptible to attack by sodium magnesium sulfate [35].

Also somewhat dated, but of interest, are the views of Mather. He said that when the Bureau of Reclamation recommendations are observed, that is, Type V cement is used, 'no significant deterioration of concrete due to sulphate attack has been encountered' [36].

An enormous contribution in the field of sulfate attack on concrete was made by Thorvaldson in the 1950s [37]. (In 1962, I had the privilege of knowing him, and I profited from his wealth of knowledge of cement.) He confirmed the influence of the type of cement on the resistance to sulfate attack of concrete with a high w/c. As usual, the studies were comparative with respect to Type I cement. However, Type I cement has changed considerably since then, mainly by way of an increase in the ratio of C_3S to C_2S. In consequence, the products of hydration contain more calcium hydroxide than is generated by C_2S. This makes the cement more vulnerable to sulfate attack with the consequence of the formation of gypsum. Al-Amoudi *et al.* observed that, in the case of magnesium sulfate, a lower content of calcium hydroxide in hydrated cement is undesirable because it encourages the reaction of sulfate with C-S-H, leading to the softening of the matrix, mass loss, and a reduction in strength [38].

The positive influence of a low C_3S to C_2S ratio was reported also by Irassar *et al.* [39]. In my view, their paper is an example of an investigation involving a single set of conditions, and great care is required before generalizations can be made. At least on evidence to date, we should refrain from saying that a low C_3S to C_2S ratio is a significant positive factor in the choice of cement for good sulfate resistance.

Furthermore, consideration of the ratio of C_3S to C_2S is often not included in the interpretation of laboratory tests; in the case of concrete in situ, the ratio is usually unknown.

Nowadays, in addition to Portland cement, there are commonly used various cementitious materials, such as fly ash and ground granulated blastfurnace slag (slag). They inevitably affect the chemical reactions involving sulfates. I am not considering these cementitious materials because the situation when only Portland cement is used should be disentangled first. With additional cementitious materials, the complexity of sulfate attack becomes even greater; for example, Taylor [24] pointed out that, if slag has a low alumina content, it improves the sulfate resistance, but with a high content of alumina, the reverse is the case [24]. On the other hand, if calcium hydroxide becomes bound by, say, fly ash, the sulfate resistance may improve.

A further complication arises from a possible interaction between the type of cement and the w/c. This makes it difficult to interpret test results in which one parameter at a time is varied. To illustrate the situation I shall refer to the work of Stark; he conducted long-term studies, and reported his findings in 1989 [40] and again in 2002 [41].

According to Stark's first report [40], visual assessment of concrete specimens half-submerged in a sodium sulfate-rich soil with periodic wetting and drying, showed that Type V cement performed better than Type I cement at values of w/c between 0.75 and 0.45; at $w/c = 0.36$, the two cements were equally good. To me, these results show a superior performance of Type V cement, and not a predominant influence of w/c. A generalization is, however, not possible. It is worth noting that Stark commented that 'wetting and drying with crystallization of salts may have contributed to deterioration' [40].

Stark's second report [41], dated 2002, also relies on visual assessment of deterioration. He found that, at a w/c of 0.40, concrete, partly immersed in sulfate-laden soil and alternately wetted and dried, performed significantly worse when Type I cement was used than with Type V or Type II cement [41]. The w/c had a significant effect on the performance above a w/c of 0.50. At a w/c of 0.65, all cements led to a poor performance but, at intermediate values of w/c, Stark found the cement composition to be 'of some importance' [41]. Interestingly, laboratory concrete immersed continuously in a solution of sodium sulfate of the same concentration as in the field tests (65 000 ppm of sulfate), showed uniformly excellent performance regardless of the type of cement, w/c, or additional cementitious materials and surface treatments [41]. Stark's conclusion is that 'the traditional explanation of expansion of concrete due to chemical reaction of sulfate ion with aluminate-bearing cement hydration products was of relatively minor significance' [41]. He concluded that a major mechanism of deterioration in the field tests (but absent in the laboratory) is cooling and heating, and wetting and drying (by natural variation) with cyclic crystallization of sodium sulfate salts [41].

According to the classification that I have used in this section, the above mechanism should be considered to be physical attack. Much more importantly, Stark's tests show that predicting the behaviour in the field on the basis of laboratory tests under singular, uniform, and strictly controlled conditions is likely to be unreliable. This explains why various laboratory tests lead to different conclusions. Hence, my plea for a new approach to the study of sulfate attack by way of *field* performance.

Actually, varying one parameter at a time in a concrete mix is not possible. For example, at a given workability, a change in the w/c entails a change in the cement content.

Importance of w/c

Occasionally, especially in disputes, strong emphasis is placed on a low value of w/c being the essential requirement for sulfate resistance, the type of cement being considered irrelevant, or at best secondary. Other people express the contrary view. Each proponent adduces evidence supporting his or her opinion. Such a situation is highly conducive to the continuation of the present confusion, but it is not surprising.

It is not surprising because much depends on the cation in the attacking sulfate, as shown earlier in this section. There are other important factors at play as well.

There is, however, an overriding requirement, necessary but not sufficient, for good sulfate resistance of concrete, and that is its density in the sense of a very low permeability. This is achieved by good proportioning of the mix ingredients, such that full compaction (consolidation) can be obtained by actual achievement of full compaction, and by effective curing to maximize the degree of hydration of the cement paste. These requirements are obvious to an engineer but often not fully considered by laboratory experimenters. Hence, my repeated plea for consideration of field concrete.

BRE Special Digest 1 states: 'Poorly compacted concrete will be particularly vulnerable to chemical attack' [42]. Taylor says that 'the effects (of sulfate) are minimized in a dense concrete of low permeability and by using a sulphate-resisting cement, in which there is little or no aluminate phase' [32].

Consideration of low permeability to ensure durability is included in BS 5328, which says that permeability is 'governed by the constituents, their proportions and the procedures used in making concrete' [43]. Eight factors influencing durability are listed, of which one is the type of cement, and another the cement content and free w/c taken together [43].

Permeability was also considered in the tests of Khatri *et al.* [44], who found that expansion of mortar specimens immersed in a 5% sodium sulfate solution was affected *both* by the type of binder and by permeability.

With reference to the influence of w/c on resistance of concrete to external attack in general, it is useful to quote Shah *et al.* [45]: 'a relatively low w/c in cement paste does not necessarily mean a reduced permeability because of the significant influence of aggregates'. Specifically, they consider early-age cracking to be a factor in ensuring durability. The relevance to the theme of this section is that absolutist statements about w/c being the governing factor ensuring resistance to sulfate attack are not correct.

Young *et al.* [46] say that sulfate attack can be prevented by any one of three factors: a low w/c, a low calcium hydroxide content, or a low C_3A content. If their claim is valid, then adequate protection of concrete should be ensured by the use of a low w/c alone *or* by the use of Type V cement alone. In my opinion, these solutions are too sweeping and not valid under all circumstances.

Somewhat surprisingly, it has been found in several investigations, as reported by Al-Amoudi [38], that lowering w/c has a deleterious effect on the resistance of concrete exposed to magnesium sulfate, the use of

Type V cement being of no avail. A likely explanation is that, at low values of w/c, there is limited pore space to accommodate the products of reactions with sulfate, namely magnesium silicate hydrate (which has no adhesive properties) and gypsum. Care is required in translating these observations into practical recommendations about mix selection. My purpose in reporting the above findings is to show, once again, that a systematic review of the entire problem of reactions of sulfates with hydrated cement for the purpose of guidance about making durable concrete is still to come. Specifically, we often fail to differentiate between the cations in guidance documents or codes, or when applying such documents.

Bureau of Reclamation tests

The Bureau has played a pre-eminent early role in studying sulfate attack on concrete and in developing sulfate-resisting concrete. An extensive and independent long-term investigation of the behaviour of concrete exposed to sodium sulfate was started by the Bureau of Reclamation in the 1950s. The specimens were continuously exposed to a sulfate over a period of more than 40 years. In addition, the Bureau conducted accelerated tests. Unfortunately, because of funding changes, the Bureau did not analyse and report the long-term test results.

The spate of litigation involving alleged sulfate attack on concrete led several people to visit the Bureau test results and to publish their interpretations. The steady-exposure, that is, non-accelerated tests, were analyzed by Kurtis *et al.* [47]. It is not my intention to comment in detail on their analysis, but I believe the following observations are in order. First, the extensive development of what they call 'empirical model to predict concrete expansion' in their paper was revoked by an erratum published eight months later [47], and equations with new shapes were introduced. This is a great pity. Less hurried research would serve us better.

Another aspect of the interpretation by Kurtis *et al.* [47] of the test data of the Bureau of Reclamation deserves mention. The interpretation distinguishes cements with a low C_3A content (described as <8%) and cements with a high C_3A content (described as >10%). This gives the impression that the Bureau tested cements with quite a range of C_3A contents. It is true that the range was wide, but the distribution of C_3A contents was uneven and the division at 10% is fortuitous: there were no cements with a C_3A content between 8.0

and 10.0. Moreover, the distribution of values was not uniform: 81 cements with less than 8.0%, and 33 with more than 10.0%; of these 33 cements, 16 had a C_3A content of between 16.01% and 17%. Also, of the 81 cements with the C_3A content of less than 8.0%, 43 had a content higher than 6.01%, and 53 higher than 5.01%. The relevance of the value of 5.01% will be mentioned in the next paragraph.

My remarks are not meant to criticize the findings of Kurtis *et al.* [47], because it was not they who chose the cements (and the Bureau of Reclamation had its own objectives) but their interpretation in qualitative terms, such as 'low C_3A content' and 'high C_3A content' may not be justified. Even more importantly, we should note that ASTM C 150-2002 limits the C_3A content of Type V cement to 5% and also restricts the sum of the content to C_4AF plus twice the C_3A content to 25%. Thus, as I see it, conclusions in terms of high- and low-C_3A contents need careful qualification.

I have a similar reservation with respect to the use of w/c as a parameter. Whereas the value of w/c used by the Bureau ranged from 0.35 to 0.75, of the 114 mixes used, 85 had w/c between 0.46 and 0.55. Thus the real spread of w/c was minimal. This is relevant to the conclusions about the significance of w/c with respect to the sulfate resistance of concrete.

Monteiro and Kurtis (who were the co-authors of [47]) wrote another paper discussing the same test results [48]. Although they emphasize the importance of the influence of w/c on sulfate resistance of concrete, this is not entirely borne out by the test results. Their conclusion that the 'Time to failure, as measured by expansion, decreases with increasing w/c and C_3A content' is correct but it is far too broad for practical application. At one extreme, they report that the concrete that failed first (after $3\frac{1}{2}$ years) had a w/c of 0.48 [48]. This concrete was made with a cement that contained as much as 73.7% of C_3S, so the C_3S content may be significant. At the other extreme, of the nine concretes that failed in less than 25 years and had a w/c >0.55, seven had a C_3A content above 6%. Moreover, a graphical interpretation of the tests, reported by Monteiro and Kurtis [48] shows that, for values of w/c between 0.45 and 0.55, the time to failure ranged from $3\frac{1}{2}$ years (but mostly >12 years) up to more than 40 years. My view of their analysis is that it does not enable us to draw any practical conclusions, especially because only the expansion of specimens was used as a measure of sulfate attack and only exposure to sodium sulfate was tested.

In view of this, it is surprising that they report [48] a reliability analysis (of which they were co-authors) that is said to have shown 'that the effect of w/c on expansion was one order of magnitude higher than the effect of C_3A content for cements with a C_3A content of less than 8%' [49]. That phrase is followed by a parenthetical qualification: '(for cements with low C_3A)'; they mean less than 8% and not a Type V cement, whose upper limit is 5%. Strangely enough, I cannot find the exact wording attributed to Corr *et al.* in [49].

It could be thought that a person who uses (or misuses) the Bureau of Reclamation test data to 'prove' that w/c is the sole factor governing the sulfate resistance of concrete 'doth protest too much'.

The test results of the *accelerated* tests conducted by the Bureau of Reclamation starting in the early 1950s over a period of 18 years were analysed by Ficcadenti [50]. The tests used soaking the concrete specimens alternately in a sodium sulfate solution and drying in hot air; the expansion of the specimens was measured. What is significant is his finding that the expansion does not progress linearly but slows down or ceases when C_3A has been used up. Ficcadenti, therefore, concluded that the C_3A content of the cement is the crucial factor, and w/c has little influence on expansion over long periods [50].

I do not have enough information to try to reconcile the findings of Ficcadenti with those of Kurtis *et al.* [47], or alternatively to deduce the correct interpretation of the Bureau of Reclamation tests; nor is such a resolution of conflicting conclusions within the ambit of this section. My purpose in mentioning the varying interpretations of the work of the Bureau of Reclamation is to illustrate the confused and confusing state of our knowledge of sulfate attack on concrete. Possibly, someone will undertake an independent retrospective study of all of the Bureau test results.

Part III: Assessment of sulfate attack

In this part, I discuss the approach of various codes to the classification of the severity of exposure conditions prior to construction. This is followed by consideration of tests to determine the damage to concrete in situ. The Appendix to this section contains selected tables from various sources giving a classification of exposure to sulfates.

Severity of exposure to sulfates

The preceding discussion of sulfate attack was in qualitative terms, but for practical purposes we need to relate the degree of severity of field exposure to the extent of the attack on concrete. In the American and British standards (until 2003) there exist two methods of characterization of the exposure: sulfate content in the groundwater that is likely to be in contact with the concrete, and the sulfate content in the soil that will be in contact with concrete. Two general comments are pertinent.

First, how comparable are the severities of exposure when determined on water and on soil? This question will be dealt with later and so will the related issue of the extraction ratio from the soil.

Second, in all cases, the quality of sulfate is expressed in parts of SO_4 in mg per kilogram of water, that is in ppm (parts per million), generally without consideration of the type of cation in the sulfate. And yet, we have seen that different cations may result in different forms of attack. Incidentally, some earlier standards expressed the sulfate as SO_3. While it is easy to convert the amount of SO_3 into SO_4, care is required when using different documents.

British Code of Practice for structural use of concrete, BS 8110:1985 [51] recognizes four classes of exposure, plus a class (less that 300 ppm in groundwater or 1000 ppm in a 2:1 water-soil extract) that has no requirements with respect to the concrete mix. In 1997, there was published British Standard BS 5328 on selection of mix proportions [43]. This standard recognizes magnesium sulfate as separate criterion of severity of exposure.

BS 5328 has now been withdrawn and the current British Standard is the European Standard EN 206-1 [52]. With respect to sulfates, expressed as SO_4, three classes of severity are recognized (the harmless condition not being considered). It is interesting to note that the classification by exposure classes and the limiting values in EN 206-1 are not recommended for use in the UK. Instead, a complementary British Standard to EN 206-1, known as BS 8500-1, has been published. One shortcoming of EN 206-1 is that it uses the total sulfate in the soil (obtained by hydrochloric acid extraction) but allows water extraction if local experience is available. In the UK, between 1939 and 1968, the sulfate content in the soil used to be determined by acid extraction (although this was not explicitly stated) and this greatly overestimated the severity of exposure.

In the USA, ACI Building Code 318 [53] recognizes three classes of exposure in terms of SO_4 (plus the harmless condition) and prescribes a maximum value of the water/cementitious material ratio, the minimum compressive strength, and the type of cement. The latter is selected on the basis of the C_3A content. The Commentary on ACI 318 lists 'other requirements for durable concrete exposed to concentrations of sulfate': these include adequate compaction, uniformity, and sufficient moist curing [53]. The importance of compaction and curing was emphasized earlier in this section.

In addition to design codes and standards, in the UK there exist so-called Digests, which are non-mandatory guidance documents, perhaps like the ACI Guides. Guides relating to sulfate resistance of concrete have existed for many years, but were revised from time to time. A selection of extracts from various American, British, European and Canadian sources giving a classification of the severity of exposure to sulfates is given in the Appendix to this section.

In 2002, there was published a new Building Research Establishment (BRE) Special Digest 1 [42], dealing with concrete in aggressive ground. It is a long and complex four-part document, which provides design guidance for various types of buildings. Part 1 introduces six classes of aggressive chemical environment in terms of soluble sulfate, magnesium, potential sulfate from oxidation of pyrite (not considered in the present section), pH, and mobility of groundwater [54].

The class of aggressive chemical environment is combined with the 'structural performance level', which is categorized according to the service life, criticality of use, and structural details. This useful innovation distinguishes three levels: (i) structures expected to have a life of less than 30 years and unreinforced house foundations; (ii) structures with an expected life of 30 to 100 years; and (iii) structures expected to have a life in excess of 100 years and critical parts of structures, such as slender elements and hinges. This approach represents a great improvement on the traditional approach in existing codes, which use a 'one-size-fits-all' approach.

The combination of the class of aggressive chemical environment with the structural performance level determines the so-called design class. There are 13 design classes, each of which can be combined with three structural performance levels, and three cross-section thicknesses. Hence, the number of 'additional protective measures' required is obtained; these include surface protection, a sacrificial layer, and use of controlled permeability formwork. Finally, this information leads to mixes specified in terms of type of cement (or cement group, in European parlance), minimum cement content and maximum w/c.

It is useful to note that the size of the cross-section of the concrete member is taken into consideration in determining the structural performance level. This means that this approach implicitly accepts that some attack on the surface of concrete can be tolerated in thicker sections [54]. However, fine-tuning according to the thickness of individual members on a single site is, in my opinion, impractical in terms of mix delivery on site.

Consideration of mobility of groundwater in an explicit manner, rather than casually, and advice on the determination of mobility of the water, introduced in Special Digest 1 [42], are a very useful innovation.

As I mentioned earlier, BRE Special Digest 1 [42] is not a mandatory document but is intended to serve as a guide. It is thus a competent source of advice for a competent engineer, but it should not be used as a rulebook to be followed more or less blindly.

Those concerned with investigations of allegedly damaged concrete should note that the design of structures using Part 3 of the Special Digest 1 'is not intended as a basis for assessing the risk of sulfate attack to existing properties' [42].

The system devised in the BRE Special Digest 1 [42] is sound and admirable. Alas, in my view, it looks like a product of an extensive desk study, with little consideration of the realities on site, where often there does not exist a single set of physical conditions. I shall discuss the variability of site conditions later on.

I realize that it is not easy to write a comprehensive alternative set of recommendations for the choice of the concrete mixes to ensure resistance to sulfates, but I suspect that what is proposed in BRE Special Digest 1 is unlikely to find favour with designers. The choice of a mix is on the basis of 16 classes of aggressive environment, for three structural performance levels, and three thicknesses of the concrete member, compounded by 'additional protective measures' that raise or lower the quality of the concrete mix. All this is a true *embarras de richesse*. Giving a choice of all the 124 single malt whiskies at the same time might be too much for a regular drinker.

What should be of greater concern are the class boundaries of the various exposure conditions. In a sense, they are arbitrary because they have not been calibrated by measurement of recorded incidence of damage to concrete caused by sulfate attack. This, of course, would not be easy because of the variability of conditions in the field. So, although we are not sure about the exact position of class boundaries, the approach currently used is qualitatively satisfactory. However, notice should be taken of the predominant cation in the sulfate because this is relevant to the efficacy of using a Type V cement.

Extraction of sulfate from soil

The various codes classify the degree of exposure of concrete to sulfates according to the sulfate content in groundwater or in an extract from the soil. Presumably, it is intended that either of these determinations should result in the same class of severity of exposure for the purpose of mix selection.

The reason for the dual approach is that where a borehole does not contain static or flowing groundwater, it is necessary to resort to testing the soil. Where there is a choice, it is water that should be tested; this is explicitly recommended by BRE Special Digest 1 [42].

Why, then, am I discussing the two test methods? Whereas determining the sulfate content in water is a matter of a simple chemical procedure, in the case of an extract, the outcome of the same procedure will be affected by the method of obtaining the extract. The method is by no means standardized or uniform in various codes.

I should like to add that I am discussing the determination of the content of water-soluble sulfates, and not acid-soluble sulfates. The latter method is used in some codes for certain purposes, but I view it as inappropriate because the attack of concrete, if any, occurs by water in contact with its surface penetrating into the interior of the concrete; on the other hand, acids may dissolve sulfates that have only a low solubility in water.

The extract should be described by mass of water per unit mass of soil; this varies from $1:1$ to $20:1$. I understand that the California Department of Transportation [55] uses an extraction ratio of $3:1$, and the US Bureau of Reclamation and the US Army Corps of Engineers, $5:1$. The significance of the extraction is that, under some circumstances, a high extraction ratio will give a higher sulfate content, and this will lead the designer to choose a concrete mix with superior properties than otherwise would be the case.

Many earlier standards were silent on the extraction ratio. In consequence, and not surprisingly, when a party interested in demonstrating a high sulfate content was in control, it used a high ratio, sometimes as high as $20:1$, whereas the opposing party would use a very low ratio, possibly down to $1:1$. The recent Special Digest 1 prescribes a $2:1$ extraction ratio [42]. This has been used in the UK since 1975, and replaces the $1:1$ ratio used between 1968 and 1975. The change was made because, with the $1:1$ ratio, some clays yielded an insufficient amount of extract water for chemical tests; this problem, however, could be remedied by taking a larger sample of soil.

The significance of the extraction ratio depends on the type of sulfate prevalent in a given situation. Specifically, if sodium and magnesium sulfates are dominant, and calcium sulfate is a minor component, then a low extraction ratio may be appropriate because both sodium and magnesium sulfate have a high solubility, so that even a small amount of water will result in an appropriate measure of sulfate exposure. On the other hand, with calcium sulfate, a low extraction ratio may lead to a measured value of sulfate content corresponding to the maximum solubility of calcium sulfate, whereas the actual sulfate content may be much higher. This situation obtains when the sulfate content is expressed in mg of sulfate per kg of soil.

To give an example: assume that the content of calcium sulfate in the soil is such that there are 10 000 mg of sulfate per 1 kg of soil. Now, the low solubility of calcium sulfate means that a maximum of 1440 of sulfate (SO_4) can be dissolved in 1 kg of water. Thus, a 1 : 1 extract would indicate 1440 ppm, which would be reported as 1440 mg per kg of soil, even though the actual content is 10 000 mg per kg. Now, using an extraction ratio of 20 : 1, the amount of calcium sulfate would be 10 000 ppm or 10 000 mg per kg, which is the actual value in the soil because all the sulfate in the soil would be dissolved without reaching saturation of the water.

I hesitate to offer an opinion on the choice of an appropriate extraction ratio or on the class of boundaries for the severity of exposure to sulfates. However, I find the argument for expressing the limits in terms of sulfate in an extract solution, and not as sulfate in the soil, to be compelling. Thus, the European approach is preferable to the ACI method.

I believe ACI is considering a change to the method of expressing the sulfate content in soil. One proposal is to use the term 'sulfate activity index': this term is uninformative. Another proposal is to use the term 'sulfate in pore water of saturated soil': this is cumbersome and may mislead some people who associate the term 'pore water' with water in the hydrated cement paste. Moreover, the soil tested may not always have pores in it saturated, and the extract may contain more water than could fill the pores in the soil in situ. My support, for whatever it is worth, is for 'sulfate in extract solution', with the extraction ratio uniquely defined.

The difficulty of establishing the sulfate content of the soil is illustrated by the data of Marchand *et al.* [56] who compared the values obtained by extraction using the California Transportation Procedure 417 (Method A) [55] with tests on centrifuged liquid from the soil, followed by ion chromatography to yield the value of sulfate (Method B). Marchand *et al.* present a plot relating these two sets of test results, but make no statement about confidence limits. By eye, there seems to be a large scatter; for example, here are four pairs of results (in ppm of SO$_4$):

| Method A: | 7000 | 7000 | 10 000 | 9000 |
| Method B: | 5000 | 11 000 | 8000 | 11 000 |

Unless a statistically sound explanation of these results is available, we should beware of making definite claims about relationships. Rather, we should establish a reliable and universally acceptable method of determining the sulfate content of the soil.

Finally, there exist some newer chemical methods, such as inductively coupled plasma atomic spectroscopy, and they may not yield the same results as other methods, but this is for chemists to say.

Taking soil samples

At the beginning of Section 4.3, I admitted that even though I am not a chemist, I felt justified in writing about chemical attack on concrete. At this stage, I would like to write about some soil aspects of sulfate attack on concrete, even though I am not a geotechnical engineer. (Of course, these topics were included in my civil engineering curriculum.) Why? The answer is that both chemical and geotechnical considerations are provided for engineering purposes; construction is pre-eminently an engineering matter.

In documents dealing with sulfate attack on concrete, there is generally no guidance on taking soil samples. There is a recent exception to this, namely the BRE Special Digest 1 [42], which gives advice on taking samples and on the interpretation of the test results on sulfates in water and in the soil. Generally, the distribution of sulfates in soil varies both horizontally and vertically. In the UK, often the top metre or so of soil is relatively free from sulfates because of leaching by groundwater [35]. On the other hand, in dry regions, where the rate of evaporation is high, there may be a higher concentration near the surface [35].

In my, admittedly limited, experience of large sites for multiple buildings construction, I learned that there may be a considerable variability in the different factors involved. For example, if the site is sloping, the conditions of groundwater may vary from place to place. They will vary also depending on whether sampling was done just after a heavy rainfall or after a prolonged period of drought. The conditions can also vary on a given site in consequence of excavation elsewhere. The combinations are numerous. And yet, for practical purposes, the foundations for a number of, say, houses, cannot be finely tuned to the conditions established at every individual site. Moreover, the measurements may well be taken prior to exact siting of every individual house.

At the same time, it would be prohibitively expensive to go on testing until the 'worst spot' has been found and to make the conditions there govern the entire site. Nevertheless, such an extreme approach has been taken; it was said that, if one part of the land has a high sulfate content and the rest only a low or even negligible content, then the high content governs. The opinion expressed was: 'take enough results and you'll find that it (high sulfate content) exists throughout'. In other words, all the foundations have to be designed for the most severe condition encountered. Accepting such reasoning leads to demonstrably nonsensical consequences.

Let us say that we propose to construct 100 individual homes. Let us assume that there is an adjacent existing development on a soil in which a high sulfate content was established some time ago. Given that the soil and sulfates are not aware of ownership boundaries, a high sulfate content in the soil on an adjacent property means, according to the 'logic' quoted above, that all 100 proposed homes need to have foundations that will resist the high sulfate content on the adjacent property. It follows further that no testing for sulfate is necessary. This saves money that would otherwise be used for testing but a great deal of money *may* be wasted by an unnecessary provision for resistance to a high sulfate content in the soil. Indulging in *reductio ad absurdum*, the conditions on a single site could be extrapolated to be valid further and further away.

Some help in solving this dilemma is offered by Special Digest 1 [42], which recommends calculating a 'characteristic' value, rather than taking the maximum value, except when only a small number of soil samples has been taken. When there are 5 to 9 results, the mean of the two highest values is taken as the characteristic value; and when there are more than 10 results, the mean of the highest 20% is used [42].

This advice is helpful, but it is less clear what to do on very large sites, such as those for, say, 100 individual houses. Much depends on the uniformity of soil in geological terms and on the extent of disturbance and grading that the soil has been subjected to. I am not proffering advice because this problem is more within the province of soil or geotechnical engineers.

Because water is essential for sulfates to interact with concrete, the possible movement of water should be considered in the construction of foundations. When these are shallow, isolation of concrete from the ambient material is possible. Alternatively, or additionally, drainage around the foundations can be provided. Similarly, a means of carrying rainwater from the roof away from the foundations should always be provided, even if the climate is thought to be dry.

I am sure I am not alone in having observed a drainage system vitiated by the house owner who extended the garden right up to the house wall and then provided continual watering so as to make the concrete almost permanently wet. The situation can be aggravated by the use of fertilizers for plants in close proximity to foundations.

Many fertilizers contain potassium sulfate and magnesium sulfate. I understand that the so-called greening agents, which impart a vivid green colour to a lawn, contain ammonium sulfate. This salt is known to be particularly aggressive to concrete. The relevance of the above is that gardening procedures, including lawn and garden irrigation, can greatly alter the sulfate conditions that existed at the time of construction. Now, changes in the conditions in service are beyond the scope of those involved in construction. Even if instruction handbooks and injunctions given to homeowners do not allow the irrigation water to run off onto foundations, such an approach is usually ineffective. Either a complete moisture barrier around the foundations needs to be provided (and this is not cheap or easy) or the homeowner has to bear the consequences of his or her actions.

At the other extreme, once the soil has been compacted, there may be limited availability of water, so that some sulfates in the soil will not be available for reaction with hydrated cement paste. According to Pye and Harrison, 'a dry site where the depth of the water table is difficult to find in any season is unlikely to give rise to significant chemical attack on concrete placed on it' [57].

My tests

I am not a newcomer to the study of sulfate attack on concrete. In 1969, I published a paper about my tests on the behaviour of concrete in saturated and weak solutions of magnesium sulfate [29]; the period of exposure was 1000 days. What is relevant to this section is the type of tests that I performed to assess the deterioration of concrete. These were: a change in the weight of the specimen; a change in the dynamic modulus of elasticity; and a change in the length of the specimen. I did not use a determination of the compressive strength of concrete because this is a destructive test so that, at each age, a different specimen would have been tested, and having many dozen specimens for a single mix would have been impractical. Compressive strength can be determined on cores when the health of a structure at a particular moment is at issue.

In my 1969 paper, I noted that 'the changes in length, weight, and frequency do not all show the same tendency, as the influence of reactions of corrosion (sulfate attack) on the three properties of concrete varies' [29]. This begs the question: which test is the most appropriate?

The answer is that much depends on the mechanism involved in the attack, which in turn depends on the cation of the attacking sulfate. Calcium sulfate can result in the formation of ettringite, which may (but not necessarily so) lead to expansion. This can occur with sodium sulfate as well, but sodium sulfate also leaches calcium hydroxide, and this would result in a decrease in the dynamic modulus of elasticity and possibly also in a decrease in weight.

I have couched the last paragraph in rather tentative terms because, in a sulfate solution, just as much as in water, cement continues to hydrate, and this initially improves the properties of the specimen. It is only in the longer term that the sulfate reactions dominate over the consequence of continuing hydration.

Now, magnesium sulfate has a deleterious effect on various properties of concrete and specifically on compressive strength, in consequence of decomposition of C-S-H. However, if magnesium hydroxide (brucite) is allowed to remain undisturbed on the surface of the concrete specimen, a protective layer will be formed and the sulfate attack will be self-limiting.

In view of the above, I have to admit that my tests were not structured as well as I would do now, but then hindsight has a 20/20 vision. The difference in the action of various sulfates should be borne in mind; however, on site, it is also possible for more than one type of sulfate to be involved at the same time.

Change in compressive strength

If there is a dispute about whether or not concrete in a structure has suffered damage due to sulfate attack, testing may be necessary. It is well known that the presence of voids in concrete decreases its compressive strength, the rule of thumb being that 1% voids decreases the strength by about 5.5%. It is, therefore, logical to expect leaching of products of hydration or the formation of cracks to cause a reduction in the compressive strength. Thus, if cores taken from an existing structure have a lower strength than the 28-day strength at the time of placing the concrete, then it is reasonable to conclude that the concrete has suffered damage, possibly owing to sulfate attack.

This approach has been used on a number of occasions. A few examples should suffice. Ibrahim *et al.* wrote 'The effectiveness of these materials (surface treatment) in decreasing the sulfate attack was evaluated by measuring the reduction in compressive strength' [58]. Harrison [5] determined the changes in compressive strength to establish whether sulfate attack had taken place in 15-year-long field exposure tests. These findings were cited with approval by Figg [59]. In his laboratory studies, Brown [60] also used the loss of compressive strength of mortar specimens to assess the damage by sulfates. Al-Amoudi *et al.* [61] have also used the change in compressive strength in their study of the performance of blended cements exposed to magnesium and sodium sulfates.

I would like to emphasize that what was measured in the references reviewed above is the change in the compressive strength consequent upon sulfate attack (or only exposure to sulfates) and not the level of strength as such, which depends on the mix used. I subscribe to the view that determining the *change* in compressive strength is one means of detecting sulfate attack; the absence a reduction in strength indicates that there has been no attack, even if some reactions have taken place. It is unfortunate that I have been wrongly listed as a reference in support of the view that the determination of compressive strength of cores is irrelevant to proving that sulfate attack has or has not taken place [62]. I am also misrepresented in the quotation from my paper on high-alumina cement concrete by the words: 'strength, especially compressive strength, is an inappropriate measure of durability' [6]. The common first author of these two references is Skalny.

I am citing the above spurious reliance on my publications because the difference between the level of strength and the change in strength is vital. What is significant with reference to sulfate attack is that an increase in strength (except at very early ages) is inconsistent with such an attack, that is, with deterioration of concrete by the action of sulfates.

Laboratory tests on concrete exposed to sulfates are generally performed on unstressed specimens whereas, in some service situations, stress may act concurrently with sulfate attack. Laboratory tests on concrete under stress have shown that 'the durability of concrete under sodium sulfate attack is affected by the state and level of sustained stress' [63]. However, the stress–strength ratio applied was generally higher than would be the case in practice. Nevertheless, the effect of stress on damage during exposure to sulfates represents an additional complication in the study of sulfate attack.

'Layered damage' and tensile strength

In the absence of documents and clear-cut procedures for the assessment of damage caused by sulfate attack, there is a danger of one-off methods, sometimes quite fanciful, being used to prove one's point. Two of these may be worth mentioning: the use of tensile strength (to the exclusion of compressive strength) and demonstration (real or alleged) of so-called layered damage.

It is to be expected that, when an attack on concrete takes place from one face of a concrete element, the extent of damage will be progressive with the distance from the exposed surface. This, for example, is observed in the case of carbonation.

A much more complex situation has, however, been proposed by some investigators who have introduced the concept of 'layered damage', which implies that concrete consists of a number of layers, parallel to the exposed surface, these layers having distinct properties. In consequence, if a tensile force is applied at right angles to the layers, failure will occur at a low tensile strength, while a compressive force in the same direction would not be affected by the layered nature of the damaged concrete.

A 'proof' of the existence of layered damage was provided by Ju *et al.* [64] by his measurement of ultrasonic pulse velocity in different layers of cores taken from a concrete slab. I have reviewed that work in a book [7] so that only a brief mention here should suffice. The pulse velocity measurements were taken at levels 15 mm apart on four equally spaced diameters, and the average values of velocity at each level were reported. On applying a statistical analysis, I found that the mean value of *all* eight levels lay within 95% confidence limits for *one* level, so that the differences between the mean values at the different levels were not statistically significant; thus the presence of 'layers' was not proved. Ju has not challenged my views.

Another 'proof' of layer damage was proffered by an expert by the use of the term 'layers' by Taylor [32]. On a closer study of his book [32], I found that by the words 'surface layers may be progressively removed' he meant laboratory preparation of successive surfaces for examination by X-ray diffractometry.

The assumption of layered damage has led to the proposition that compressive strength tests are incapable of detecting sulfate damage, but that the determination of direct tensile strength does do so. I should make it clear where I stand on this issue because my studies of concretes suspected of having been subject to sulfate attack have shown that the ratio of the splitting tensile strength to the compressive strength of such concretes is the same as for concrete stored under normal conditions [7]. I should point out that I am referring to the splitting tensile strength, and not to the direct tensile strength, because the latter is not standardized by ASTM or by a British Standard, so that a reliable test method is not available; for that matter, testing concrete in direct tension is very difficult because of the danger of eccentricity and of complex end stresses.

The suggestion that the compressive strength is unaffected by sulfate attack, while the splitting tension strength is lowered, is founded upon a 1982 report by Harboe [65]. His tests were performed on cores from the spillway of a dam in Wyoming, built in the mid-1930s, the cores being tested in 1967. Harboe found the compressive strength to be 40 MPa (5900 psi) while the tensile strength was only 2.2% of that value. Normally the tensile strength would be expected to be about 10% of the compressive strength.

I have no explanation for those results but they have not been independently confirmed on any other concrete in the succeeding 35 years. Harboe himself did not offer an explanation for what appears to be an anomalous ratio of splitting tensile strength to compressive strength. I find it difficult to accept this anomaly because, if concrete has developed cracks and contains damaged and soft hydrated cement paste, how can its compressive strength remain unimpaired? Proponents of selective influence of alleged sulfate attack upon mechanical properties of concrete have offered no answer to my question about tensile and compressive strengths; my view is that both compressive and tensile strength must be affected by damage in the same manner.

In addition to quoting Harboe, proponents of testing concrete in direct tension also rely on Ju *et al.* [64] who say: 'the uniaxial compression test is *not* indicative or suitable for layer-damaged concrete cylinders' [64]. (The italics are in the original text.) Because the compression test is, according to them, not suitable, they proceed to use direct tension testing, basing themselves on ASTM D 2936-95, which is a test for 'intact rock core specimens'. Using standardized tests for one material on a different material is a dubious practice.

Further work on tensile testing was done by Boyd and Mindess [66], who interpreted their tests on concrete exposed to sulfates to show that the splitting-tensile strength is less sensitive to damage by sulfate. My

145

interpretation of their results is that they simply conform to the general pattern of variation in the ratio of the tensile strength to compressive strength; this ratio decreases with an increase in strength. Thus, I do not see their tests as supporting the hypothesis of level of tensile strength being an indicator of sulfate damage.

Moreover, Boyd and Mindess [66] introduced what they describe as a 'novel tensile strength test', actually developed in England in 1978, in which gas pressure is applied to the curved surface of a cylinder. Unfortunately, as stated by the inventor of the test, N. Clayton [67], the value of strength so determined varies according to the choice of loading medium. Perhaps this is why the 'new' method died a death, and resurrecting it 20 years later does not seem to be profitable.

Locher [68] is cited as having used, as far back as 1966, the tensile strength to determine sulfate resistance of mixes containing slag. This is correct in that he tested $10 \times 10 \times 60$ mm mortar prisms exposed to various conditions [68]. However, given the shape and size of his specimens, the determination of compressive strength was impossible. There is no indication in his paper that he considered the measurement of flexural strength preferable or superior to the compressive strength; this was simply a necessity. Thus, I see reliance on Locher as a proponent of testing mortar (and presumably concrete) in tension, as against testing in compression, to be unfounded.

Use of scanning electron microscopy

This section is not concerned with advanced testing techniques, but a brief comment on the use of scanning electron microscopy (*sem*) may be in order. In some cases, *sem* has been used to establish that sulfate attack had occurred by demonstrating the presence of ettringite. As I understand the situation, *sem* does not show the presence of compounds, but only of chemical elements, so that compounds are inferred from the quantities of elements that would be present in the given compound. So the presence of, say, ettringite cannot be unequivocally proven.

Moreover, the mere presence of ettringite is not a proof of damage because much depends on whether the ettringite is located in pre-existing voids or cracks, or whether the cracks were the consequence of expansion by ettringite formation. Furthermore, Detwiler *et al.* say that 'ettringite is not a sign of sulfate attack unless its concentration is greater than would be expected from hydration of cement' [69]. It is not easy to establish how much ettringite originates from the hydration of cement because generally the composition of the cement is not known at the time of the investigation, which is usually several years after construction. Moreover, Detwiler *et al.* point out that, 'in concrete exposed to conditions of wetting and drying, it is possible that sulfate concentration would be higher just below the surface than immediately at the surface due to washing out of ions' [69]. This is highly relevant to the fact that *sem* 'looks' at an exceedingly small surface. Typically, using a 1000-fold magnification, the area whose elemental analysis is determined would be 10^{-6} m by 10^{-6} m and about 2×10^{-6} m deep. For comparison, human hair has a typical diameter of about 80×10^{-6} m (but there is a decreasing availability of hirsute men).

Because of the small size of the *sem* picture, Detwiler *et al.* recommend the following procedure in the investigation of suspected sulfate attack: 'Step-by-step progression from large scale (tens of metres) to submicroscopic scale (tens of nanometres] gave context to the testing, allowing for selection of specimens that suitably represented the concrete as a whole and for appropriate interpretation of results' [69]. I find this to be a logical approach.

There is also an important influence of relative humidity to which the test specimen is exposed in *sem* (except in the latest generation of instruments). This problem has been very clearly considered by Kurtis *et al.* [70]. They point out that 'Characterization techniques that require high vacuum or drying... are not particularly appropriate as artifacts are introduced' and go on to say that 'methods, such as scanning electron microscopy... have given limited and misleading information about expansion reactions in concrete ...' [70]. Accordingly, they recommend that use of soft X-ray transmission microscopy, which does not require drying or application of pressure [70]. Enthusiasts of scanning electron microscopy might like to take note.

The danger of artifacts created by a high vacuum in the sample chamber is pointed out also by Zelić *et al.* [71].

Mathematical modelling

A totally different approach to assessing the damage by sulfates is by the use of mathematical models. They are likely to become a valuable tool but until they have been calibrated and confirmed by field behaviour – and we are not there yet – they will not provide an understanding of the physical phenomena involved. In addition, the boundary conditions applied in a model may be arbitrary and conveniently simplistic.

147

This can happen with laboratory testing as well. Idorn criticizes laboratory procedures that confine the energy conversion to occur at a constant temperature for reasons of convenience, thus violating the true kinetics of the reactions instead of modelling the reality of concrete [72].

It is pertinent to quote a statement made in the *Magazine of Concrete Research* about site tests (for a different purpose): 'Conditions on site are not controlled in the same way as in the laboratory, but the benefits of assessing real effects at full scale are obvious, and allow greater confidence in modelling and prediction procedures derived from data presented and, by extrapolation, from data obtained by others' [73]. It is also said about site data: 'The benefits of being able to confirm predictions made on the basis of short-term or small-scale experiments are evident' [73].

What I find highly surprising [64] and sometimes even at a date preceding the construction of a building. And yet, the building is still in use, and has neither been condemned nor recommended for urgent repairs. As an engineer, I find a theory that is flatly contradicted by reality too difficult to accept.

A problem with models is the choice of boundary conditions, which sometimes may distort the output of a model. For example, one theoretical analysis concluded that 'numerical simulations indicate that the exposure to weak sodium sulfate solutions may yield (sic) to a significant reorganization of the internal microstructure of concrete' [74]. This may be true but it does not necessarily mean sulfate *attack* and damage to concrete. Support for my statement is given by DePuy (referred to in the same paper): 'Chemical reactions *per se* in concrete are not necessarily harmful' [1].

Sometimes the authors developing theoretical analyses cite numerous references to support their assumptions, but such support is diminished when the references are to tests on concrete mixes of a bygone era, which used cements significantly different from those used nowadays. To give one example, the statement that: 'There is overwhelming evidence to show that the degradation also contributes to a significant reduction in the mechanical properties of concrete' [74] cites: a paper from the 1920s, another from the 1940s, two papers on laboratory tests, including neat cement paste specimens, and a book by the proponent of the theoretical analysis in question [74]. I expect that the use of 'self-support' is not all that uncommon: after all, every one of us has confidence in our own publications. However, there is the danger that a statement about 'overwhelming evidence' is likely to be cited again and may well become embedded in the received wisdom, but is it really correct?

Another example of a sweeping generalization is a statement that 'none of them (building codes) contains any specific limit concerning the maximum w/c that should be selected for the production of concrete elements to be exposed to negligible levels of sulfate. This is unfortunate since, as emphasized by many authors, potentially destructive conditions may exist even though analyses indicate the groundwater or soil to have a low sulfate content' [74]. The references supporting the preceding statement include a paper published in 1968, another published in 1982 and a 1994 paper.

I sometimes wonder whether some of the papers discussing concrete in Southern California alleged to have been attacked by sulfates have not been written to establish the damage of those concretes. The models then developed are premised on sulfate attack that had not been physically established in an independent manner. I am not imputing ill will because we can all be subject to what Bertrand Russell, a great philosopher, called 'the delusive support of subjective certainty'.

There is a more frivolous, but nonetheless valid, support of my view that an opinion repeated often enough, even if incorrect, acquires currency. Lewis Carroll (a Cambridge mathematician, best known as the author of *Alice in Wonderland*) wrote in *The Hunting of the Snark*, published in 1876: 'What I tell you three times is true'.

Part IV: An overview

Conclusions
Amid the plethora of confusing information and opinion, some general conclusions can be drawn.

The term 'attack' should be used only to describe damage to concrete, and not just the fact that reactions have taken place. Furthermore, the term 'sulfate attack' should be applied only to the consequences of chemical reactions involving the sulfate ion, recognizing that all reactions have physical consequences. Damage by sulfates where the action is purely physical and which can also be produced by other salts should be termed 'physical attack'.

Sulfate attack on concrete structures in service is not widespread except in some areas, and the amount of laboratory based research seems to be disproportionately large. On the other hand, our knowledge and understanding of sulfate attack in the *field* remains inadequate.

The efficacy of Type V cement as against the use of a low w/c in reducing sulfate attack on concrete is still unresolved, but there is no doubt that dense concrete, well compacted (consolidated), and well cured is essential.

The significance of the cation in the sulfate is often not appreciated, and yet this influences the chemical reactions and thus the consequences of sulfate attack.

The classification of the severity of exposure of concrete to sulfates is not firmly established and has not been calibrated by the determination of actual damage to concrete under different conditions. A review of classifications used in the past is given in [75].

The classification of exposure conditions should be clear but also simple. It is not sensible to make elaborate and complex recommendations for the selection of the concrete mix when this classification is predicated on largely uncertain and variable measurements of sulfate in the ground. We are not dealing with an industrial product whose property can be measured precisely, but rather with a variable natural material – soil – that is furthermore disturbed by human operations during construction and by the user of the structure who may change the drainage or introduce salts while beautifying or maintaining the land around the building.

Drainage of soil around the foundations is important because dry salts do not attack the concrete.

The classification of severity by the determination of the sulfate content in the soil is still a matter of debate. The determination of the sulfate content in the soil should be standardized in terms of the extraction ratio, so that the outcome of the test cannot be varied according to the whim or objectives of the tester. The amount of sulfate should be expressed in mg of SO_4 per kg of extract water.

Assessment of the extent of damage by the determination of the *change* in compressive strength appears to be reliable. On the other hand, the argument that only tensile strength testing should be used seems to be unfounded.

Mathematical models will be of help in understanding sulfate attack, but, so far, they have not been verified.

The use of scanning electron microscopy is not a primary evidence of sulfate attack but should be preceded by tests on a large scale.

What next?

I would also like to draw some broader inferences. Alas, I believe that the title of this paper is correct: our knowledge of sulfate attack is confused, and the practitioners in construction lack a clear picture that would enable them to achieve sulfate-resisting structures. Individual researchers know much about specific actions and materials, but the ensemble of their knowledge does not hold together. Unfortunately, a reconciliation of these various opinions is not possible at the present state of our knowledge. What this paper has tried to achieve is to point out the conflicting views, with reasons for the divergences in opinion where these could be ascertained. Thus, we know, at least *why* the world of sulfate attack is confused.

The purpose of this section was to present, as objectively as I could, the problems in assessing the sulfate attack on concrete, and to point out the ramifications of generalized statements about damage to concrete. To resolve the various issues much needs to be done. In other words, we need research.

But what research? Limiting oneself to laboratory tests is not enough. It is not particularly helpful to use accelerated tests on miniscule specimens of neat cement paste or mortar. Realistic conditions of exposure need to be achieved.

I would like to think that this section will be of use in showing where *practical* research is needed and is likely to pay off. But this research has to be related to *field* conditions of exposure to sulfates.

To say that researchers should be objective is to state the obvious. But with large-scale litigation being the driving force, all too often research is undertaken to 'prove a point' for one party or another, or to secure a 'victory'. Once a researcher has become bound to his or her paymaster, even if he or she wishes to remain objective, the researcher moves towards a goal and narrows the horizon of his or her work. Moreover, even if this is not the case, he or she is perceived to be biased, and this prompts others working on the topic to prove the researcher wrong. And so it goes on.

I am not saying the above to criticize anybody because we are all fallible 'sinners'. What we have to beware of, however, is that sweeping assertions that are advanced, without objective and verifiable documentation available for all to see, are not good science. These assertions should not become a part of the body of knowledge upon which all concrete specialists in the world at large can rely.

I am not recommending a mammoth experimental investigation. I believe that much can be gleaned from re-visiting published information both on laboratory test results and on site observations, as well as from tests on field concrete. This needs to be done by a team of chemists, petrographers, and structural engineers all working together.

Thomas Macaulay, an early 19th century historian, wrote in *How Horatius Kept the Bridge*: 'Now who will stand on either hand and keep the bridge with me'.

The leader of the team should be a *new* person versed in matters structural, and he or she should be flanked by someone (on the right hand) versed in chemistry of concrete, not just cement, and by someone (on the left hand) experienced in optical microscopy of field concrete. If eyebrows are raised at the inclusion of structural engineers, my answer is that the objective of establishing facts about sulfate attack and, hence, providing guidance on achieving sulfate-resisting *structures*, is to build structures that will be durable for the required service life under the expected conditions of exposure.

I have used the term 'sulfate-resisting structures' rather than sulfate-resisting concrete because, all too often, the field of vision of researchers is limited to the material, whereas the purpose of studying concrete is to build good and durable structures. If we do not succeed in this, competitors to concrete structures – and we must not delude ourselves by thinking that concrete is irreplaceable – will gain the ascendancy. But then I am a structural engineer and not a backroom researcher; I offer no apology for this.

At the outset of this paper, I said that sulfate has generally not caused significant damage to structures in service. So why this major discussion and recommendations for an independent review and study? The answer is that a lack of clarity permits some people to claim sulfate attack and, at present, there is no reliable and coherent body of knowledge that would make it easy to distinguish real damage from spurious allegations. The reaction of some designers has already been to over-provide protective measures so as to avoid potential (even if unjustified) claims in the future. This costs the client money and reduces the competitiveness of concrete.

It would be unsatisfactory for this section to end by observing simply that our knowledge of sulfate attack on structures in service is confusing. On the other hand, we are not yet in a position to present a coherent and complete picture on the basis of which we could select suitable concrete mixes and design durable structures, as well as assess objectively the state and probable future of existing structures. To achieve this, we need to review all the available information and to pull it together, clearly distinguishing laboratory results and field observations. It is the latter – structural behaviour – that is of paramount importance, but scientific knowledge is also necessary.

Finally, I would like to make a very general remark: there is an incompatibility between the breadth of specified limits on the properties of cement on the one hand and, on the other, the more demanding expectations of concrete, which can be achieved by the use of cementitious materials with specific properties. Now, Portland cement needs to comply only with very broad limits on strength, fineness, and chemical compounds. At the same time, requirements, such as compatibility with superplasticizers or ensuring sulfate resistance, need close and narrow controls. Whenever this issue is raised, cement manufacturers reply that you cannot have a high-tech material for the price of a low-tech one. Which do we want? Consideration of this dilemma requires a separate study.

An essential part of sulfate is sulfur, and the element sulfur has long been known to contribute to damage. In *King Lear*, Shakespeare says:

'There's hell, there's darkness, there is the sulphurous pit,
   Burning, scalding, stench, consumption; fie, fie, fie!'
   ......

'Singe my white head!'

which, given my white head, is an appropriate ending to this section.

References

1. DePuy, G.W., Chemical resistance of concrete, in *Concrete and Concrete-Making Materials*, ASTM STP 169C, Philadelphia, PA, 1994, pp. 263–281.
2. Hime, W.G., Chemists should be studying chemical attack on concrete, *Concrete International*, 25, 4, 2003, 82–84.
3. Crammond, N., The occurrence of thaumasite in modern construction: a review, *Cement and Concrete Composites*, 24, 3–4, 2002, 393–402.
4. Bensted, J., Thaumasite – three-year review, *Concrete*, Sept., 2003, 7.
5. Harrison, W.H., *Sulphate Resistance of Buried Concrete*, Building Research Establishment Report, 1992.
6. Skalny, J.P., Odler, I. and Marchand, J., *Sulfate Attack on Concrete*, Spon, London and New York, 2001.
7. Neville, A., *Neville on Concrete*, ACI, Farmington Hills, MI, 2003.
8. Mehta, P.K., Sulfate attack on concrete: separating myths from reality, *Concrete International*, 22, 8, 2000, 57–61.
9. The Concrete Society, *Diagnosis of Deterioration in Concrete Structures*, Technical Report, No. 54, 2000.
10. Collepardi, M., A state-of-the art review on delayed ettringite attack on concrete, *Cement and Concrete Composites*, 25, 2003, 401–407.
11. Skalny, J., Odler, I. and Young, F., Discussion of [12], *Cement and Concrete Research*, 30, 2000, 161–162.

12. Hime, W.G. and Mather, B., 'Sulfate attack,' or is it?, *Cement and Concrete Research*, 29, 1999, 789–791.

13. Lafuma, H., Théorie de l'expansion des liants hydrauliques, *La revue de matériaux de construction et de travaux publics*, Part 1, 243, Dec., 1929, 441–444 and Part 2, 244, Jan., 1930, 4–8.

14. Bonnell, D.G.R. and Nottage, M.E., Studies in porous materials with special reference to building materials. I. The crystallization of salts in porous materials, *Journal Society Chemical Industry*, 58, 1939, 16–21.

15. Lea, F.M. and Nurse, R.W., Problems of crystal growth in building materials, *Discussions of the Faraday Society*, 5, 1949, 345–351.

16. Brown, P.W. and Taylor, H.F.W., The role of ettringite in external sulfate attack, in *Sulfate Attack Mechanisms, Materials Science of Concrete*, American Ceramic Society, Ohio, 1999, pp. 73–97.

17. Mehta, P.K., *Discussion, in Sulfate Attack Mechanisms, Materials Science of Concrete*, American Ceramic Society, Ohio, 1999, pp. 22–23.

18. Flatt, R.J., Salt damage in porous materials: how high supersaturations are generated, *Journal of Crystal Growth*, 242, 2002, 435–454.

19. Brown, P.W., Thaumasite formation and other forms of sulfate attack: Guest editorial, *Cement and Concrete Composites*, 24, 3–4, 2002, 301–303.

20. Neville, A., Concrete: 40 years of progress?, *Concrete*, 38, Feb., 2004, 52–54.

21. Mehta, P.K., Sulfate attack on concrete: a critical review, in *Materials Science of Concrete III*, American Ceramic Society, Westerville, OH, 1993, pp. 105–130.

22. Pierce, J.S., Discussion, in *Sulfate Attack Mechanisms, Materials Science of Concrete*, American Ceramic Society, Westerville, OH, 1999, pp. 41–43.

23. Tobin, R.E., *Sulfate Reactions in Concrete, Important Technical Information*, May, 22, 2000.

24. Taylor, H.F.W., Discussion, in *Sulfate Attack Mechanisms, Materials Science of Concrete*, American Ceramic Society, Westerville, OH, 1999, pp. 33–34.

25. Santhanam, M., Cohen, M.D. and Olek, J., Sulfate attack research – whither now? *Cement and Concrete Research*, 31, 2001, 845–851.

26. Tian, B. and Cohen, M.D., Expansion of alite paste caused by gypsum formation during sulfate attack, *Journal of Materials in Civil Engineering*, 12, 1, 2000, 24–25.

27. Mather, B., Discussion of the process of sulfate attack on cement mortars by Shen Yang, Xu Zhongzi, and Tang Mingshu, *Advanced Cement Based Materials*, 1997, 109–111.

28. Rasheeduzzafar, O.S.B., Al-Amoudi, S.N., Abduljauwad, M. and Maslehuddin, M., Magnesium-sodium sulfate attack in plain and blended cements, *ASCE Journal of Materials in Civil Engineering*, 6, 2, 1994, 201–222.

29. Neville, A.M., Behavior of concrete in saturated and weak solutions of magnesium sulphate or calcium chloride, *ASTM Journal of Materials*, 4, 4, 1969, 781–816.

30. Shah, V.N. and Hookham, C.J., Long-term aging of light water reactor concrete containments, *Nuclear Engineering and Design*, 185, 1998, 51–81.
31. Eglinton, M., Resistance of concrete to destructive agencies, in *Lea's Chemistry of Cement*, Fourth Edition, Arnold, London, 1998, pp. 299–342.
32. Taylor, H.F.W., *Cement Chemistry*, Second Edition, Thomas Telford, London, 1997.
33. Bureau of Reclamation, *Concrete Manual*, Eighth Edition, US Government Printing Office, Washington, DC, 1985.
34. Harrison, T., The influence of cement content on the performance of concrete, *Concrete Society Discussion Document*, CS 125, 1999.
35. Lea, F.M., *The Chemistry of Cement and Concrete*, Third Edition, Arnold, London, 1970.
36. Mather, B., Field and laboratory studies of the sulphate resistance of concrete, in *Performance of Concrete*, University of Toronto Press, Toronto, Canada, 1968.
37. Thorvaldson, T., Chemical aspects of the durability of cement products, *Third International Symposium on Chemistry of Cement*, London, England, 1952, pp. 436–484.
38. Al-Amoudi, O.S.B., Attack on plain and blended cements exposed to aggressive sulfate environments, *Cement and Concrete Composites*, 24, 3–4, 2002, 305–316.
39. Irassar, E.F., González, M. and Rahhal, V., Sulphate resistance of Type V cements with limestone filler and natural pozzolana, *Cement and Concrete Composites*, 22, 2000, 361–368.
40. Stark, D., *Durability of Concrete in Sulfate-Rich Soils*, Portland Cement Assn, Research and Development Bulletin, RD O97, Skokie, IL, 1989.
41. Stark, D., *Performance of Concrete in Sulfate Environments*, Portland Cement Assn, Research and Development Bulletin, RD 129, Skokie, IL, 2002.
42. BRE Special Digest 1, *Concrete in Aggressive Ground*, Parts 1–4, Second Edition, BRE Centre for Concrete Construction and Centre for Ground Engineering Remediation, 2003.
43. British Standard BS 5328: Pt 1, *Specifying Concrete*, BSI, London, 1997.
44. Khatri, R.P., Sirivivatnanon, V. and Yang, J.L., Role of permeability in sulphate attack, *Cement and Concrete Research*, 27, 8, 1997, 1179–1189.
45. Shah, S.P., Wang, K., and Weiss, W.J., Mix proportioning for durable concrete, *Concrete International*, 12, 9, 2000, 73–78.
46. Young, J.F., Mindess, S., Gray, R.J. and Bentur, A., *The Science and Technology of Civil Engineering Materials*, Prentice-Hall, Upper Saddle River, NJ, 1998.
47. Kurtis, K.E., Monteiro, P.J.M. and Madanat, S.M., Empirical models to predict concrete expansion caused by sulfate attack, *ACI Materials Journal*, 97, 2, 2000, 156–167, and Erratum 97, 6, 2000, 713.

48. Monteiro, P.J.M. and Kurtis, K.E., Time to failure for concrete exposed to severe sulfate attack, *Cement and Concrete Research*, 33, 7, 2003, 987–993.

49. Monteiro, P.J.M., Kurtis, K.E. and Der Kiureghian, A., Sulfate attack of concrete: reliability analysis, *ACI Materials Journal*, 98, 2, 2001, 99–104.

50. Ficcadenti, S.J., Effects of cement type and water to cement ratio on concrete expansion caused by sulfate attack, *Structural Engineering, Mechanics and Computation*, 2, 2001, 1607–1613.

51. British Standard BS 8110, *Structural Use of Concrete*, BSI, London, 1985.

52. European Standard EN 206-1, *Concrete – Specification, Performance, Production and Conformity*, 2000.

53. ACI Building Code, 318, 2002.

54. Nixon, P.C., Concrete in aggressive ground: an introduction to BRE Special Digest 1, *Concrete*, Sept., 2001, 62–64.

55. California Department of Transportation, Test 417, *Method of Testing Soils and Waters for Sulfate Content*, 1986.

56. Marchand, J., Samson, E., Maltais, Y., Lee, R.J. and Sahu, S., Predicting the performance of concrete structures exposed to chemically aggressive environment-field validation, *Materials and Structures*, 35, Dec., 2002, 623–631.

57. Pye, P.W. and Harrison, H.W., Performance, diagnosis, maintenance, repair and avoidance of defects, in *BRE Building Elements: Floors and Flooring*, Construction Research Communications Ltd, London, 1997.

58. Ibrahim, M., Al-Gahtani, A.S., Maslehuddin, M. and Dakhil, F.H., Use of surface treatment materials to improve concrete durability, *ASCE Journal of Materials in Civil Engineering*, 11, 1, 1999, 36–40.

59. Figg, J., Field studies of sulfate attack on concrete, in *Sulfate Attack Mechanisms, Materials Science of Concrete*, American Ceramic Society, Ohio, 1999, pp. 315–323.

60. Brown, P.W., An evaluation of the sulfate resistance of cements, *Cement and Concrete Research*, 11, 1981, 719–727.

61. Al-Amoudi, O.S.B., Maslehuddin, M. and Saadi, M.M., Effect of magnesium sulfate and sodium sulfate on the durability performance of plain and blended cements, *ACI Materials Journal*, 92, 1, 1995, 15–24.

62. Skalny, J. and Pierce, J.S., Sulfate attack: an overview, in *Sulfate Attack Mechanisms, Materials Science of Concrete*, American Ceramic Society, Ohio, 1999, pp. 49–63.

63. Schneider, U. and Piasta, W.G., The behaviour of concrete under Na_2SO_4 solution attack and sustained compression or bending, *Magazine of Concrete Research*, 43, 157, 1991, 281–289.

64. Ju, J.W., Weng, L.S., Mindess, S. and Boyd, A.J., Damage assessment of service life prediction of concrete subject to sulfate attack, in *Sulfate Attack Mechanisms, Materials Science of Concrete*, American Ceramic Society, Ohio, 1999, pp. 265–282.

65. Harboe, E.M., Longtime studies and field experiences with sulfate attack, in *George Verbeck Symposium on Sulfate Resistance of Concrete*, SP-77, ACI, Farmington Hills, MI, 1982, pp. 1–18.

66. Boyd, A.J. and Mindess, S., The effect of sulfate attack on the tensile to compressive strength ratio of concrete, in *Third International Conference on Concrete Under Severe Conditions*, ACI/CSCE, Vancouver, 2001, pp. 789–796.

67. Clayton, N., Fluid-pressure testing of concrete cylinders, *Magazine of Concrete Research*, 30, 102, 1978, 26–30.

68. Locher, F.W., Zur Frage des Sulfatwiderstands von Hüttenzementen (The problem of the sulfate resistance of slag cements), *Zement-Kalk-Gips*, 9, 1966, 395–401.

69. Detwiler, R.J., Taylor, P.C., Powers, L.J., Corley, W.G., Delles, J.B. and Johnson, B.R., Assessment of concrete in sulfate soils, *ASCE Journal of Performance of Constructed Facilities*, 14, 3, 2000, 89–96.

70. Kurtis, K.E., Monteiro, P.J.M., Brown, J.I. and Meyer-Ilse, W., Expansive reactions in concrete observed by soft X-ray transmission microscopy, *Materials Research Society Symposium Proceedings*, 524, 1998, 3–9.

71. Zelić, J., Krstulović, R., Tkalčec, E. and Krolo, P., Reply to the discussion of the paper 'Durability of the hydrated limestone-silica fume Portland cement mortars under sulphate attack', *Cement and Concrete Research*, 30, 2000, 333.

72. Idorn, G.M., Sulphate attack on concrete, in *Sulfate Attack Mechanisms, Materials Science of Concrete*, American Ceramic Society, Upper Saddle River, OH, 1999, pp. 65–72.

73. Editorial, *Magazine of Concrete Research*, 40, 145, 1988, p. 194.

74. Samson, E., Maltais, Y., Beaudoin, J.J. and Marchand, J., Theoretical analysis of the effect of weak sodium sulfate solutions on the durability of concrete, *Cement and Concrete Composites*, 24, 3–4, 2002, 317–329.

75. Longworth, I., Development of guidance on classification of sulfate-bearing ground for concrete, *Concrete*, 38, Feb., 2004, 25–26.

Appendix: Classification of exposure to sulfates

The Appendix contains a selection of tables from American, Canadian, British and European sources giving a classification of exposure to sulfates.

These tables show that it is not only the sulfate content boundaries of the various exposure classes that vary, but also the soil–water extraction ratios (sometimes unspecified), as well as the method of determining the sulfate content: acid or water soluble. Furthermore, the concentration of the sulfate in the soil is sometimes expressed per unit mass of soil; in other cases, per unit mass of extraction water. The rationale of these various approaches is not self-evident.

Even a quick look at the ensemble of the classifications of exposure conditions and at the class boundaries shows substantial differences. However, their full significance can be assessed only by a consideration of the properties of mixes corresponding to the various exposure classes.

Such a consideration is not possible because the properties of the various cements were not the same in the different countries. Moreover, the actual properties of cements varied over time, and the standard requirements for cements were also not immutable.

I am not saying that such a study is impossible, but it would be a very major undertaking. Moreover, the study would be meaningful only if we had a record of long-term performance of the various concretes under the actual exposure conditions. Alas, with very few exceptions of long-term field studies, one in California and one in England, no such information is available.

The preceding discussion exemplifies our dilemma: we do not know whether the exposure class boundaries are 'correct' because they have not been calibrated or verified. This then is the reason for the section entitled, 'The confused world of sulfate attack on concrete' and this is why the entire arena of sulfate attack in *the field* needs to be studied.

Table 4A.1 ACI 2001.2R (1997) Guide to Durable Concrete; *ACI 332R (1984, reapproved 1997)* Guide to Residential Concrete

Class of exposure	Concentration of sulfates as SO_4	
	In groundwater: ppm	Water-soluble in soil: %
Mild (0)	<150	<0.10
Moderate (1)	150–1500	0.10–0.20
Severe (2)	1500–10 000	0.20–2.00
Very severe (3)	>10 000	>2.00

Table 4A.2 ACI 318 (1983 through 2002) Building Code

Class of exposure	Concentration of sulfates as SO_4	
	In water: ppm	Water-soluble in soil: % by mass
Negligible	<150	<0.10
Moderate	150–1500	0.10–0.20
Severe	1500–10 000	0.20–2.00
Very severe	>10 000	>2.00

Table 4A.3 US Bureau of Reclamation Concrete Manual *(1966)*

Class of exposure	Concentration of sulfates as SO_4	
	In water: ppm	Water-soluble in soil: %
Negligible	0–150	<1000
Positive	150–1000	1000–2000
Considerable	1000–2000	2000–5000
Severe	>2000	>5000

159

Table 4A.4 Uniform Building Code (1991)

	Concentration of water soluble SO_4	
Class of exposure	In groundwater: ppm	In soil: %
Negligible	<150	<0.10
Moderate	150–1500	0.10–0.20
Severe	1150–10 000	0.20–2.00
Very severe	>10 000	>2.00

Table 4A.5 Canadian Standard A23.1.94

	Concentration of sulfates as SO_4	
Class of exposure	In groundwater: ppm	Water-soluble in soil: %
S–3 moderate	150–1500	0.10–0.20
S–2 severe	1500–10 000	0.20–2.0
S–1 very severe	>10 000	>2.0

Table 4A.6 European Standard EN 206-1:2000

Class of exposure (chemical environment)	Concentration of sulfates as SO_4		Concentration of magnesium in groundwater: ppm
	In groundwater: ppm	In soil: ppm total	
Slightly aggressive	200–600	2000–3000	300–1000
Moderately aggresive	600–3000	3000–12 000	1000–3000
Highly aggressive	3000–6000	12 000–24 000	3000 to saturation

Notes: 1. The extraction of SO_4 from soil is by hydrochloric acid. However, water extraction may be used if experience is available in the place of use of the concrete.
2. Clay soils with a permeability below 10^{-5} m/s may be moved into a lower class.

Table 4A.7 *British Building Research Station: Notes on concrete in sulfate-bearing clays and ground waters, The Builder, 7 July 1939, p. 29. This table was incorporated in British BRE Digest 31 (1951)*

	Concentration of sulfates as SO_3	
Class of exposure	In water: ppm	In clay: %
1	<300	<0.2
2	300–1000	0.2–0.5
3	>1000	>0.5

Notes: 1. Analysis of groundwater is preferred to that of clay.
2. In some cases, it may be found that the results of the groundwater and clay analyses lead to different classifications; the more severe classification should be adopted.

Table 4A.8 *British Code of Practice CP 110:1972*

	Concentration of sulfates as SO_3		
Class of exposure	In groundwater: ppm	In 2:1 water–soil extract: gram per litre	Total: ppm
1	<300	–	<2000
2	300–1200	–	2000–5000
3	1200–2500	1.9–3.1	5000–10 000
4	2500–5000	3.1–5.6	10 000–20 000
5	>5000	>5.6	>20 000

Notes: 1. For Classes 3 to 5, if calcium sulfate is predominant, the water–soil extract may give a lower class of exposure than the total SO_3.
2. To convert SO_3 to SO_4, multiply by 1.2.

Table 4A.9 British Code of Practice CP 8110:1985

Class of exposure	Concentration of sulfates as SO_3		
	In groundwater: gram per litre	In 2:1 water–soil extract: gram per litre	Total: ppm
1	<0.3	<1.0	<2000
2	0.3–1.2	1.0–1.9	2000–5000
3	1.2–2.5	1.9–3.1	5000–10 000
4	2.5–5.0	3.1–5.6	10 000–20 000
5	>5.0	>5.6	>20 000

Notes: 1. If calcium sulfate is predominant, the water–soil extract will give a lower class than total SO_3.
2. To convert SO_3 to SO_4, multiply by 1.2.

Table 4A.10 British Standard BS 5328:1997 Guide to Specifying Concrete

Class of exposure	Concentration of sulfate as SO_4			Concentration of magnesium	
		In soil			
	In groundwater: ppm	By acid extraction: %	By 2:1 water–soil extract: gram per litre	In groundwater: ppm	In soil: gram per litre
1	<400	<0.24	<1.2	–	–
2	400–1400	*	1.2–2.3	–	–
3	1500–3000	*	2.4–3.7	–	–
4A	3100–6000	*	3.8–6.7	≤1000	≤1.2
4B	3100–6000	*	3.8–6.7	>1000	>1.2
5A	>6000	*	>6.7	≤1000	≤1.2
5B	>6000	*	>6.7	>1000	>1.2

Notes: 1. Consider both SO_4 and magnesium.
2. Classification on the basis of groundwater is preferred.
3. Higher values are given for the water–soil extract in recognition of the difficulty of obtaining representative samples and of achieving a comparable extraction rate to that from groundwater.
* Classify on the basis of 2:1 water–soil extract.

Table 4A.11 British BRE Digest 90 (1968)

| Class of exposure | Concentration of sulfates as SO_3 | | |
	In groundwater: ppm	In 2 : 1 water–soil extract: gram per litre	Total: ppm
1	<300	–	<2000
2	300–1200	–	2000–5000
3	1200–2500	2.5–5.0	5000–10 000
4	2500–5000	5.0–10.0	10 000–20 000
5	>5000	>10.0	>20 000

Note: To convert SO_3 to SO_4, multiply by 1.2.

Table 4A.12 British BRE Digest 174 (1975)

| Class of exposure | Concentration of sulfates as SO_3 | | |
	In groundwater: ppm	In 2 : 1 water–soil extract: gram per litre	Total: ppm
1	<300	–	<2000
2	300–1200	–	2000–5000
3	1200–2500	1.9–3.1	5000–10 000
4	2500–5000	3.1–5.6	10 000–20 000
5	>5000	>5.6	>20 000

Note: To convert SO_3 to SO_4, multiply by 1.2.

Table 4A.13 British BRE Digest 250 (1981)

| Class of exposure | Concentration of sulfates as SO_3 | | |
	In groundwater: ppm	In 2 : 1 water–soil extract: gram per litre	Total: ppm
1	<300	<1.0	<2000
2	300–1200	1.0–1.9	2000–5000
3	1200–2500	1.9–3.1	5000–10 000
4	2500–5000	3.1–5.6	10 000–20 000
5	>5000	>5.6	>20 000

Note: To convert SO_3 to SO_4, multiply by 1.2.

Table 4A.14 British BRE Digest 363 (1991)

| Class of exposure | Concentration of sulfates as SO_3 | | | Concentration of magnesium | |
| | In groundwater: ppm | In soil | | In groundwater: ppm | In soil: gram per litre |
		By acid extraction: %	By 2 : 1 water–soil extract: gram per litre		
1	<400	<0.24	<1.2		
2	400–1400	*	1.2–2.3		
3	1400–3000	*	2.3–3.7		
4a	3000–6000	*	3.7–6.7	<1.0	<1.2
4b	3000–6000	*	3.7–6.7	>1.0	>1.2
5a	>6000	*	>6.7	<1.0	<1.2
5b	>6000	*	>6.7	>1.0	>1.2

* If >0.24, classify on the basis of 2 : 1 water-soil extract.

Table 4A.15 British BRE Special Digest 1 (2002) British Standard BS 8500:2000

| Class of exposure | Concentration of sulfate as SO_4 | | Concentration of magnesium | |
	In groundwater: ppm	In 2 : 1 water–soil extract: gram per litre	In groundwater: ppm	In 2 : 1 water–soil extract: ppm
DS–1	<400	<1200		
DS–2	400–1400	1200–2300		
DS–3	1500–3000	2400–3700		
DS–4a	3100–6000	3800–6700	≤1000	≤1200
DS–4b	3100–6000	3800–6700	>1000	>1200
DS–5a	>6000	>6700	≤1000	≤1200
DS–5b	>6000	>6700	>1000	>1200

Note: 1. Consider both SO_4 and magnesium.
2. Static and mobile water are distinguished.
3. The value of pH of the water is also considered.

4.4 SULFATE IN THE SOIL AND CONCRETE FOUNDATIONS

Adam Neville and Robert E. Tobin

In August 2004, one of us (A.N.) published in the journal *Cement and Concrete Research* a long paper, reviewing research and practice, titled 'The confused world of sulfate attack on concrete' (see Section 4.4). That paper showed that our knowledge of sulfate attack on concrete in the *field* is woefully inadequate. Moreover, the various rules in codes and their interpretation, especially by lawyers, are not soundly based. Is it therefore reasonable to take legal proceedings against contractors in the absence of reliable knowledge of, for example, the minimum level of the sulfate content that can reliably be deemed to lead to attack, damage, need for repairs and a claim for compensation?

The paper provoked some discussion but, in the meantime, a new editor of the journal decided that it should concern itself solely with narrowly construed 'primary research data', rather than exhaustive reviews of past research, and refused to publish the discussion. We, on the other hand, think that the objective of research should be to achieve good concrete in *practice*. This is why we are amplifying A.N.'s article for the benefit of those who mix, sell, place and consolidate concrete, and who do not want to be hounded without reason.

One of us (R.E.T.) has 70 years' experience as a structural engineer, much of it in California. A.N. has investigated the foundations and slabs on grade of numerous homes in southern California as an expert witness. It is on this basis that we are writing this section. The main points are as follows.

With the exception of a handful of problematic homes, we found no damage caused by, or consequent upon, sulfate attack. We define sulfate attack as destructive chemical action and damage; the presence of sulfates by itself is not a proof of anything untoward as far as the concrete in the field is concerned. As late as last year, The Portland Cement Association said that 'terminology on this topic (sulfate reactions) is still being developed by the industry'.

So, the situation is that, in homes 10 to 20 years old, no structural damage has occurred and no concrete needs replacement.

Even more remarkable is what has happened in those lawsuits that were settled with payment, sometimes large, made to homeowners and, of course, to their attorneys. To our knowledge, those moneys have not been used to repair, let alone replace, the foundations or slabs on grade. The homes continue to be occupied and are fit for the purpose; in a number of cases, the property has changed hands, the purchaser seemingly not being worried about safety.

The absence of replacement or significant repairs in the 'lawsuit homes' is an indirect proof that attack and damage had not occurred: not a single letter reporting such events was written to the editor of the journal that published A.N.'s article.

The homeowners' lawyers have won some of the lawsuits because it was held by court that there was a violation of the Building Code insofar as the maximum water–cement ratio of the concrete was concerned. The applicability of some tables in the Building Code to slabs on grade and foundations is possibly a matter for discussion. We do not believe that they are applicable but an authoritative decision is needed. The ACI Building Code ACI 318 does not concern itself with slabs on grade. A.N.'s views are presented in Sections 5.3 and 5.4 in this book. It would therefore be best for the engineer in charge of a project to specify the mix. This would do away with some lawyers' arguments that the contractor or the ready-mixed-concrete supplier should ensure conformance with the design code.

One of us (R.E.T.) has observed surface sulfate incrustations on concrete in parts of California, Montana and Wyoming. The sites were in regions of low relative humidity and low rainfall. It is likely that groundwaters were drawn to the surface by evaporation. It was the natural moisture within the soil that dissolved the available sulfates. When water evaporated at the surface due to the low humidity of the air, a white deposit of sulfate minerals was left on the ground surface. (Fig. 4.4.1) These are referred to locally as 'alkali flats'. Similar behaviour can be observed in concrete on the ground without a protective plastic moisture barrier, but this is not damage.

A serious element in the present confused situation is the large gap between laboratory research and field construction. Laboratory studies very often deal with calcium sulfate as the principal salt involved, but in the field there may be other sulfates instead of, or in addition to, sodium sulfate. The action of the various sulfates is not the same. There is also a wide gap between the findings of the laboratory studies and their usefulness on the jobsite. So, the assessment of G.W. de Puy expressed in 1994 still stands: 'Although sulfate has been extensively investigated, it is still not completely understood.'

At the present time, there is no proven justification for the various values of the sulfate content in the groundwater, which is usually

Fig. 4.4.1. Incrustations of salts on the stem wall inside garage. (The stem wall is poorly compacted)

classified as mild, moderate or severe. If we do not know that the description is correct, we should not use it to select the mix.

Likewise, there is no agreed extraction ratio of sulfate from the soil sample. Each party to a dispute can choose the ratio, from 1:1 to 1:20, so as to 'prove' its point.

This situation does great harm to the satisfactory use of concrete in foundations and slabs on grade (Fig. 4.4.2). It also exposes the concrete

Fig. 4.4.2. Etching of garage floor near entrance from drive. The core shows that the concrete is good

providers, in the broad sense of the word, to unnecessary and unjustified litigation. An honest and unbiased review of field behaviour is needed without delay.

The title of A.N.'s paper used the word 'confused'. A considerable amount of verbiage has been published on the subject of sulfate reactions, but the amount of disagreement that prevails after many years of research is very considerable, and the situation can be described as 'tangled'. The Scottish writer, Sir Walter Scott, wrote, at the beginning of the 19th century: 'O what a tangled web we weave, when first we practise to deceive.' In general, researchers are not prone to deceive but their results have certainly become tangled.

4.5 BACKGROUND TO MINIMIZING ALKALI–SILICA REACTION IN CONCRETE

Damaging alkali–silica reaction (ASR) in concrete is rare in most parts of the UK. When it does occur, however, its consequences can range from aesthetic to very serious, so that we should always choose a concrete mix with a minimum risk of the occurrence of damaging ASR. Given that damaging ASR is not common, why am I writing this section?

Rationale of this section

The answer is bizarre. The main source of alkalis in concrete is the alkalis in cement (Na_2O and K_2O) whose origin is the raw materials and fuel introduced into the kiln that produces the cement clinker. In modern cement plants, the raw materials are pre-heated by the hot gases leaving the upper end of the kiln. These gases contain a significant proportion of the volatile alkalis, and a part of the gases may need to be bled off to control the alkali content of the cement [1]. This is one means of controlling the alkali content in the cement.

BS EN 197-1:2000 [2] expresses the alkali content as the percentage of equivalent soda Na_2O eq [which is (Na_2O + $0.658(K_2O)$)] in the mass of cement. Generally, three levels of alkali content in cement – low, moderate, and high – are considered in a matrix with three levels of silica reactivity of aggregate: low, normal, and high [3].

Such an approach had worked well until, on 24 January 2005, Lafarge Cement UK – probably the largest cement manufacturer in the world – issued a press release confirming that between September 2002 and December 2004 the reported values of alkali levels in the cement produced at the Westbury Works in Wiltshire had been falsified by 'a small group of rogue employees'. This letter was widely reported in the construction press [4].

If a concrete mix was designed on the assumption that, say, a moderate alkali cement was used, but it transpires later that the cement had a high alkali content, then ... back to the drawing board. But, more alkalis in cement than expected is not necessarily problematic, and this is why I am looking at the mix designs used to minimize the risk of disruptive ASR.

Conditions necessary for ASR to occur

There are three necessary conditions for ASR to proceed: alkalis, reactive silica and adequate moisture. This is usually taken as not less

Fig. 4.5.1. This section is concerned with minimizing ASR, but it is nevertheless useful to know what ASR looks like. Here is a typical situation. Cracking parallel to the exterior surface of the concrete goes both through and around the aggregate particles. Also, although it is difficult to discern, the alkali–silica gel lines the surface of some of the cracks and air voids. (Photograph taken by Margaret Hanson Reed, WJE Northbrook IL)

than 85% relative humidity [1] (which is common outdoors in the UK at night or in winter and which may be present in the interior of concrete due to residual mix water). Somewhat surprisingly, BS 8500-1:2002, clause A.12.1 says that 'these actions [given in BS 8500-2:2002 apply regardless of whether the concrete will be in a dry environment' [5].

Alkalis are always present in Portland cement, and often also in admixtures; to guarantee an absence of reaction would necessitate using aggregate containing no reactive silica. Fortunately, in the UK, aggregates which can be classified as having a normal or low alkali reactivity are generally available. If such aggregate is not available locally, importing it over a long distance can be uneconomic.

Here may be an appropriate place to mention that sodium chloride (NaCl) is a source of alkalis; for example, poorly washed sea-dredged sand may contain residual sodium chloride. Also, de-icing salts on pavement or runway contain sodium, and so may some groundwater. From stoichiometry, 1 gramme of chloride ions is equivalent to 0.76 gramme of Na_2O.

Although, in an ideal world, the risk of damaging ASR can be avoided, the approach of British Standards is to minimize the risk of damaging ASR, rather than to guarantee its absence.

Minimizing the risk of ASR

The objective of minimizing risk, rather than eliminating it totally, is not unusual in design. Indeed, the concept of minimizing the risk is embedded in British codes. For example, BS 8110:1985 Structural use of concrete says [6]: 'the aim of design is the achievement of an acceptable probability that the structures being designed will perform satisfactorily during their intended life' [6]. Clearly, 'acceptable probability' is less than 100%. As recently as April 2005, Heyman [7] wrote in *The Structural Engineer*: 'The actual load paths in the structure cannot be known, but the engineer is nevertheless required to provide a safe and practical design' [7]. And also, 'whereas the loads may be specified accurately, the boundary conditions will, in general, be unknown'.

It is not surprising, therefore, that BS 8500-1:2000, clause 4.12.1 says: 'the producer is required to take action to minimize damaging alkali–silica reaction' [5]. It follows that it is generally not reasonable for a structure owner to require a *guarantee* that ASR will not occur.

Exact meaning of the term 'alkali content in the cement'

The term 'alkali content in the cement' needs to be elaborated. According to BS EN 197-1:2000 there are two measures of alkali content in cement: 'declared mean alkali content' and 'certified average alkali content' [2].

There is also 'guaranteed alkali limit', which is used exclusively with low alkali cement, the limit being 0.60%. This limit must 'not be exceeded by any test result on any spot sample', as stipulated in BS EN 197-1:2000, clause B 4.3, Note 2 [2].

BS EN 197-1:2000, clause NB 4.3(b) says: 'A declared mean alkali content is an alkali content expressed as the sodium oxide equivalent, which must not be exceeded without prior notice from the manufacturer. It is a certified average alkali content plus a margin that reflects the manufacturer's variability of production' [2].

Note 1 says: use declared mean alkali content (rather than the certified average content) for purposes of calculation of the mass of alkalis from cement.

BS EN 197-1:2000, clause NB 4.3(c) says 'certified average alkali content is an average of the manufacturer's latest 25 consecutive determinations on spot samples, taken in accordance with a statistically based sampling plan, e.g. autocontrol' [2]. In clause NB 4.3(d) the variability of a certified average alkali content is represented by the standard deviation of the manufacturer's latest 25 consecutive determinations [2]. Now, BS EN 197-1:2000, clause 3.14 defines a spot sample as a 'sample taken at the same time and from one and the same place, relating to the intended tests. It can be obtained by combining one or more immediately consecutive increments (see EN 196-7)' [2].

More generally, the same standard defines autocontrol testing as 'continual testing by the manufacturer of cement spot samples taken at the point(s) of release from the factory/depot' [2]. For the purpose of determining the alkali content in the cement, the point of release is presumably the discharge from the cement silo into tankers. This leads to an interesting question: how are consecutive samples reckoned when a period of no sales of cement intervenes, e.g. between Christmas and New Year? Because a kiln must not stop and cool down, clinker production continues during such a period but this clinker is not ground.

The definitions of BS EN 197-1:2000 largely leave the details of the testing plan to the cement manufacturer. All that is said under Definitions is that: the sample be taken at the same time and the same place; and that the characteristic value of the property tested be satisfied [2]. Importantly, the choice of the characteristic value is left to the manufacturer and, seemingly, it need not be routinely disclosed to the purchaser of the cement.

The distinctions between the different categories of alkali content are of importance insofar as compliance with British Standards is concerned. Curiously enough, Lafarge Cement UK claims that, whereas the certified average alkali contents from the Westbury works were falsified, the determinations of the alkali content in spot samples was correct.

This brings me to the issue of autocontrol for monitoring production, a concept first introduced by the European Standards on cement. Autocontrol is applied by the manufacturer and requires the use of a statistically based plan, which is referred to in BS EN 197-1, clause NB 4.3(a) [2]. It would be interesting for engineers (and not just cement plant chemists) to know how a sampling plan is selected in a given plant and how it is applied and verified. For example, as already mentioned, the calculation of the certified average alkali content starts with 25 consecutive determinations on spot samples: in practice, how often are the spot samples taken? Is the running average examined like a cusum? Is a running standard deviation calculated? How is the variability of the standard deviation, as well as of the mean alkali content, reflected in the information supplied to the purchaser of cement? And finally, is the operation of the autocontrol ever inspected and verified?

The prefix 'auto' does not mean that the control is done by itself (as automobile means moving by itself). Rather, the word 'autocontrol' is a neologism. According to the *New Shorter Oxford Dictionary*, 1993, 'auto' is used in words from Greek or modelled on Greek with the sense 'one's own, by oneself, independent...'. In our context, I take autocontrol to mean that the control of alkali content is exercised by the cement manufacturer and not by an external body.

Any problematic situation apart, it is clear from the definition of the declared mean alkali content that we never know precisely the alkali content of any consignment of cement from a given silo. A *fortiori* we do not know the alkali of an individual batch of concrete or even of concrete from a ready-mixed plant supplied on a single day. This is not a criticism of the situation: I am describing it only to point out that we have to rely on the *correct value* of the declared mean alkali content.

If there is a small departure form this value, the consequences have to be assessed by the structural engineer for the project. The involvement of an engineer is important because he or she is used to exercising engineering judgement. I am mentioning this because I have come across a materials specialist who tended to act as a 'number cruncher' and who mindlessly compared numbers in a British Standard with calculated values derived from several materials in the mix, each including an inevitable error, and treated the comparison as a go–no-go gauge.

Usual specifications for minimizing ASR

These days, it is common to use the *National Structural Concrete Specification for Building Construction* [8] complemented by a Project Specification for the project in hand. Particularly important is confirmation from the producer of concrete that the concrete conforms to BS 8500-1:2002, clause 5.2 [5]. This requires, as appropriate, details of service record, verification of conformity, and relevant manufacturers' guaranteed alkali limit or declared alkali content. Thus, the concrete producer must provide information about the alkali content in the aggregate, generally originating from seawater on sand, and also the alkali content in admixtures. According to BS 8500-2:2002, clause C.1 [9], the latter can be established in one of three ways: either as the measured value, or as the specified maximum value given in the admixture standard, or else as the guaranteed alkali limit. Some admixtures, e.g. superplasticizers, have a considerable alkali content.

Clause 5.2 of BS 8500-2:2002 is crucial to minimizing 'damaging alkali–silica reaction' [9], but I found the rules to be complex. In essence, the content of Na_2O eq in 1 cubic metre of *concrete* from constituents *other than cement* is not to exceed one of two values: 0.20 kg or 0.60 kg.

For each of these values, there is also a limiting value of alkali content in $1 m^3$ of concrete. This limiting value is a product of the declared mean alkali content of cement and cement content in $1 m^3$ of concrete. The total alkali content in concrete arising from the sum of the alkalis in cement and in the constituents other than cement not exceeding 0.20 or 0.60 kg per $1 m^3$ leads to a minimum risk of damaging ASR.

A number of combinations giving the limiting value of the product of alkali content in cement and cement content in concrete are given in Tables 4 and 5 and BS 8500-2:2002 [9]. Here are some other examples of acceptable combinations. For a maximum of 0.20 kg of Na_2O eq in concrete from sources other than cement, the British Standard gives the declared mean alkali content in cement of 0.80% and a limiting cement content of $375 kg/m^3$; or 1.00% declared mean alkali content and a limiting cement content of $300 kg/m^3$.

For a maximum 0.60 kg of Na_2O eq in the concrete from sources other than cement, the British Standard gives the declared mean alkali content in cement of 0.80% and a limiting cement content of $300 kg/m^3$.

Numerous other combinations are possible and interpolation of tabulated values is to be used. All this is logical and offers a wide choice of mix proportions but I, for one, find the procedure cumbersome, especially as two fairly large tables are used, and each one has five footnotes. Personally, I would find it easier to follow a more direct approach. On the other hand, ready-mixed concrete suppliers may find it easy to establish a routine procedure, and I admit that the tables make for easy checking of the acceptability of the proposed mix proportions.

I should add that the presence of fly ash or ground granulated blast-furnace slag in the mix has a somewhat complicated influence on the possible development of ASR, depending on the content of these materials in the total mass of cementitious materials; in most cases, their presence is beneficial with respect to ASR.

Responsibility for minimizing ASR

This brings us to the question of whose responsibility it is to select the mix proportions so as to minimize damaging ASR. There seems to be a divergence of views between BS EN 206-1:2000, clause 5.2 [10] and what is described as 'the view within BSI'. This is referred to in BS 8500-1:2002 in clause A.12.1 [5]. Consequently, both the specifier and the producer should take a close interest in the selection of mix proportions, including the categorization of aggregate as being of low or normal reactivity: proof from the aggregate supplier may be desirable.

Although BS 8500-1:2002, clause A.12 [5] says that the requirements of BS 8500-2:2002 apply 'regardless of whether concrete will be in a dry environment', the same clause says that 'if concrete is to remain in a dry environment, the specifier may specify less exacting requirements than those given in BS 8500-2'. This somewhat flexible approach supports my view that the engineer should have a vital role in mix selection and that he or she should exercise engineering judgement.

What I have written so far shows my somewhat critical view of the indirect, if not complicated, mix design laid down in the British Standard 8500 for the purpose of minimizing the risk of ASR. Unless this approach is actually included in the Project Specification, it is possible to use the simplified guidance of BRE Digest 330 [3]. Specifically, instead of using the declared mean value of Na_2O eq content of alkalis in the actual cement used, Digest 330 recognizes three categories of cement: low alkali, moderate alkali, and high alkali. I mentioned this categorization of cement at the beginning of this section, but it is possible that the approach of Digest 330 (which is not a British Standard) may be subject to challenge, should a contractual dispute arise. In passing, I would add that low-alkali cement with an upper limit of Na_2O eq of 0.60% has been standardized and used in the USA for nearly 50 years.

Should a structure be tested periodically?

A major report of the Institution of Structural Engineers [11] published in 1992, includes an assessment of the severity rating of the risk of damaging ASR in suspect concrete: according to this risk and the

175

consequences of failure, the report recommends a pattern of periodic inspection and core testing [11]. The world has moved on since 1992, and we now have cements with a known declared mean alkali content, and we also know from experience in the UK that using properly selected concrete mixes leads only rarely to damaging ASR.

Despite the fact that damaging ASR is not commonplace in the UK, there may be cases where, through a mistake, or in consequence of deliberate falsification [4], we are confronted by a situation where the risk of damaging ASR needs to be assessed and appropriate action may be required: this includes testing.

Within the limits of the present section, I would like to mention just one approach, namely, that recommended by Fournier and Bérubé [12] for the assessment of internal damage due to ASR. Cores are tested in compression and in tension: a low ratio of the latter to the former is indicative of *existing* damage due to ASR in the interior of the concrete.

Final remarks

Minimizing damaging ASR is important, but a sense of proportion should be retained. The existing British approach has been successful in that in the last 25 years 'there have been no known cases of deterioration of structures built to these (UK guidelines)' [13]. Of course, not every ASR is damaging: sometimes, no more than exudation of alkali–silica gel occurs, and this is only of aesthetic interest, which may or may not be significant. On the other hand, if the gel, which is of the 'unlimited swelling' type is confined by the surrounding hydrated cement paste, internal pressures may arise and they may lead to expansion, cracking, and disruption of the hydrated cement [1].

Second, if concrete is thoroughly dried, ASR will stop: there will be no reversal of the reaction, but no further damage will occur until re-wetting takes place.

Third, ASR does not go on forever: if either reactant (silica or alkalis) runs out, the reaction obviously stops.

Fourth, without excessive alkali content in the mix, ASR will not occur, but the calculated values of the alkali content should not be treated as absolute boundaries.

Finally, how much actual damage occurs in consequence of a potentially damaging ASR depends on the presence of reinforcement capable of opposing the expansive forces induced by ASR. As these forces can act in any direction, generally three-dimensional reinforcement is required.

To close, I would like to quote Hill [14] on the international scene: 'There is such great diversity in natural aggregates that there is no magic number for avoiding alkali–silica reaction.'

References

1. Neville, A.M., *Properties of Concrete*, 4th Edition, Longman, London, and John Wiley, New York, 1995.
2. BS EN 197-1:2000, Cement – Part 1: composition, specifications and conformity criteria for common cements, BSI, London, 2000.
3. BRE Digest 330, *Alkali–silica reaction in concrete, Part 2: Detailed guidance for new construction*, BRE, 2004.
4. Anon., Lafarge workers falsified data on cement supplied to its ready-mixed concrete suppliers, *Construction News*, 20 January, 2005, p. 1.
5. BS 8500-1:2002, *Concrete, Complementary BS to BS EN 206-1, Method of specifying and guidance for the specifier*, BSI, London, 2002.
6. BS 8110: Part 1:1985, *Structural use of concrete. Part 1. Code of practice for design and construction*, BSI, London, 1985.
7. Heyman, J., Theoretical analysis and real-world design, *The Structural Engineer*, 19 April, 2005, pp. 14–17.
8. *National Structural Concrete Specification for Building Construction*, Third Edition, May, 2004.
9. BS 8500-2:2002, *Concrete – Complementary BS to BS EN 206-1, Specification for constituent materials and concrete*, BSI, London, 2002.
10. BS EN 206-1:2000, *Concrete – Part 1: Specification, performance, production and conformity*, BSI, London, 2000.
11. *Structural Effects of Alkali–Silica Reaction*, The Institution of Structural Engineers, Hannover, 1992.
12. Fournier, B. and Bérubé, M.-A., Alkali–aggregate reaction in concrete: a review of basic concepts and engineering implications, *Can. J. Civ. Engng*, 27, 2000, pp. 167–191.
13. BRE Info IP1/02, *Minimizing the risk of alkali–silica reaction: alternative methods*, BRE, 2000.
14. Hill, E.D., Alkali limits for prevention of alkali–silica reaction: a brief review of their development, *Cement, Concrete and Aggregates*, ASTM, 18/1, 1996, pp. 1–7.

5

Behaviour in service

This chapter deals with three specific issues: the path taken by cracks (Section 5.1), the requirements for residential slabs on grade (Sections 5.3 and 5.4), as well as with sustainability issues in general (Section 5.2).

The discussion headed 'Which way do cracks run?' (Section 5.1) arose out of a proposition that you can tell the age of cracks by establishing whether they run in a more-or-less straight line, passing indiscriminately through the matrix and the coarse aggregate or, alternatively, whether the cracks preferentially skirt around the aggregate particles. It was postulated that the choice of the path was a function of the age of the concrete when the crack developed. The purpose of the argument was to 'demonstrate' that the cracks were old and preceded a certain event, so that the financial liability for the repairs could not be linked to that event.

I felt that the proper way of assessing that proposition was to study the phenomena involved in a general way, and this is what Section 5.1 presents. By the way, a somewhat similar attempt to use cracks for the purpose of determining the age of the concrete when they opened, is discussed in Sections 4.1 and 4.2. Parenthetically, I would express the view that 'throwaway' or *ad hoc* hypotheses are not a sound scientific approach to resolving technical problems. It is preferable to study a hypothesis first and then, if appropriate, to apply it to a technical problem or to a legal dispute.

Sections 5.3 and 5.4 discuss the design code requirements for residential slabs on grade. Such slabs are very common in parts of the USA, especially in the south west, and there is little doubt that they are an economic solution under certain climatic conditions.

178

Notwithstanding the above, in the last ten years or so, there has been a spate of lawsuits about the durability of such slabs. Actually, no physical damage has been observed and no residential homes have been rendered uninhabitable or are known to have needed significant repairs. What was claimed, in terms of strict liability, is that the residential slabs had not been designed in accordance with the extant design codes.

In a way, what is at issue is the interpretation of the codes, and these have been changing in recent years. Since compliance with a code has to be established in relation to the code applicable at a specific time, it is not enough to consider only the current or most recent code. Moreover, at the beginning of this century, the Uniform Building Code was phased out and was replaced by International Codes, of which one deals specifically with residential construction. Sections 5.3 and 5.4 review the situation in detail. We can only hope that my conclusion will be accepted by the engineering and architectural community at large, and will put an end to much of the vexatious litigation. If this does not come to pass, someone should demonstrate the correct interpretation of the codes. By someone, I mean engineers or architects, and not attorneys, whose concern is advocacy rather than the establishment of a correct technical situation.

Now, Section 5.2, 'Sustainability issues', is quite different from all the other sections in this book because prior to early 2006 I had not written any paper on sustainability. Thus, I was one of the few engineers who has not gone to print on a subject which seems to have replaced motherhood as a topic of universal approbation and admiration. I know that it is verging on the foolhardy not to laud sustainability unreservedly, yet here I dare to do so.

Of course, I am not against sustainable construction, and I believe engineers have always, to a larger or smaller degree, attempted to produce designs that could be described as sustainable. In this context, by sustainable design I mean one that is economic in the use of materials and of labour, in the consumption of energy, and in operation as well as at the primary construction stage. So, for a long time, we have striven to be economical with all that goes into construction: I do not see that as being different from today's cry to leave enough materials and energy sources for our children and grandchildren. Admittedly, we were not always successful, and sometimes even outright careless, but all we need to do is to perform better.

We were certainly less good at avoiding pollution and minimizing ecological damage, but often this was simply a consequence of ignorance.

For example, we imagined that no discharge into seas or oceans could cause harm because virtually unlimited dilution is possible. Originally, we viewed rivers in the same light, but we did learn that it paid to keep rivers clean so that the water could be used, and sometimes re-used, in irrigation or for drinking.

Re-cycling of paper, glass and plastics arrived on the scene quite some time ago, but these are not, strictly speaking, within the purview of engineering sustainability. Nevertheless, there is a lesson there. Nearly 20 years ago, I heard a lecture by a professor of conservation, in which he demonstrated that re-cycling paper, other than in large conurbations, had a negative energy balance. His argument was that collecting old newspapers and transporting them over a distance of 100 or 200 miles, expended more energy than would be required to produce new newsprint. The same was the case with transporting empty bottles from the old Soviet Union to Finland for the purpose of manufacturing fibre-glass insulation, but then the Soviet Union was keen to receive payment in dollars.

It is clear from the preceding that I do not view sustainability as a sacrosanct concept but, nevertheless, I feel that I should look seriously at some aspects of sustainability: this I do in Section 5.2.

5.1   WHICH WAY DO CRACKS RUN?

Abstract
A high stress causes development of cracks, which may fracture aggregate particles or skirt around them. Numerous factors influence the path taken by a crack. No direct studies are available, but the relevant factors are: properties of the parent rock; shape, size, and texture of coarse aggregate particles; properties of the interface and bond; moduli of elasticity of the aggregate and of the matrix; and strength of the matrix relative to the strength of the aggregate. At very early ages, fracture parameters change but, later on, age is not a factor influencing the crack path. The crack path influences the shear strength of beams, in that cracks passing through the aggregate reduce the contribution of aggregate interlock: some aggregates result, therefore, in a low shear strength. This has been found only recently. Establishing experimentally the influences on the crack path would enable us to design mixes economically with balanced properties of strength and ductility.

Cracking is virtually an inherent feature of concrete, and crack widths range from submicroscopic, through microcracks, to wide cracks visible to the naked eye when the concrete is viewed at an arm's length. When testing a concrete specimen in a compression testing machine, the development of cracks can be observed and, after the test, we can view the paths that the cracks have followed. It is these paths that I wish to discuss.

What causes a crack?

I should explain at the outset that I am considering a crack caused by a single factor, that factor being a high load-induced stress, although a restrained deformation (e.g. due to temperature change) would be no different; I am not considering shrinkage cracking.

In essence, the plane of a crack is normal to the tension force causing the crack to open. This force can be direct tension, tension induced by bending, tension in a diagonal crack caused by shear in a beam or, alternatively, indirect tension induced by a compressive force in the direction of the crack. What all these cracks have in common is that, in a simple stress field of increasing intensity, a crack would widen and extend in length, approximately in the same plane or, two-dimensionally, in the same line.

In a homogeneous material, the line would be straight, or nearly so. However, in concrete, which is grossly heterogeneous owing to the

181

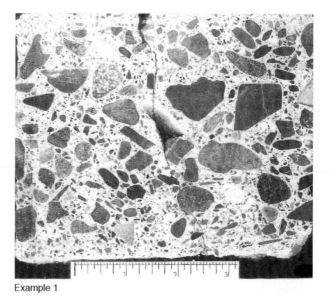

Example 1

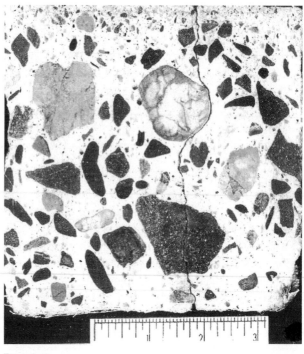

Example 2

Fig. 5.1.1. Examples of crack paths; scale in inches (courtesy of The Erlin Company, Latrobe, PA)

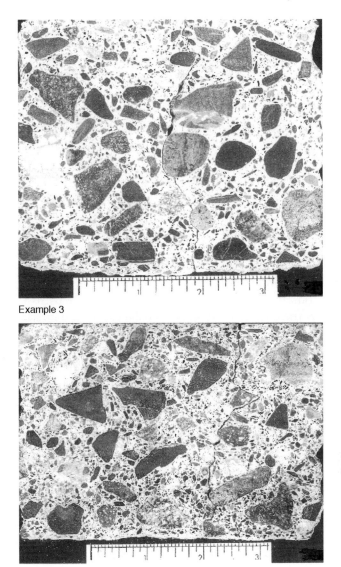

Example 3

Example 4

Fig. 5.1.1. Continued

presence of coarse aggregate particles, a straight line must travel through some of these particles. Sometimes it does, and sometimes it does not. The question is: why? Examples of crack paths are shown in Figure 5.1.1. Different mixes and different aggregates were used, and in some cases not all the aggregate particles had the same properties; the crack paths varied accordingly.

Why consider the crack path?

One reason for my interest in this topic is that I have heard it postulated that the crack path – through the coarse aggregate particles or around them – is a function of the age of the concrete when the crack opens. Specifically, it was postulated that in young concrete the cracks skirt around the aggregate and, conversely, in old concrete they pass through the aggregate particles, thus fracturing them. It was argued that the crack path would thus establish the age of the concrete at the time of cracking. This single-factor proposition led me to a literature search on factors influencing crack paths.

Of course, one could give an intuitive answer and say that the crack path is not a simple function of age. This could be the case only if there is a dramatic change in the strength of the hardened cement paste with age. This change in strength will be discussed under the heading of the 'Effect of age on strength'. At this stage, it would be sound to look at the local forces and stresses involved in crack development.

The second reason for my present interest in crack paths is that the path affects the contribution of the so-called aggregate interlock to the shear strength of concrete. As recently as 2004, it was found that, in high-strength concrete, made with limestone aggregate, 'aggregate fracturing at cracks had produced crack surfaces too smooth for the shear transfer implicit in design recommendations' [1]. This is likely to lead to an amendment to the guidance on the shear aspects of design. Hence, the topic of this section has structural significance.

Examples of crack paths are shown in Figure 5.1.1. The question about the path taken by the crack is perfectly legitimate and, indeed, interesting. In fact, it is a question that I have considered in the long-gone-by days when I instructed undergraduate students in testing concrete specimens to determine their strength. My view then was that when the hydrated cement paste (or the mortar containing that paste) is much weaker than the aggregate particles, cracks travel around the particles, leaving them intact. Conversely, when the aggregate particles are weak, they cannot act as crack arresters, and the crack develops in a straight line fracturing the aggregate particles in its path.

I have not modified my views in the intervening half-century, but I had no reason to apply myself explicitly to the issue until the present time.

The recognition of the influence upon the crack path of the strength of aggregate relative to the strength of the matrix was acknowledged by Hassanzadeh [2]. He concluded that the water–cement ratio (w/c) is therefore a factor influencing the fracture energy [2]. However, he reported that his 'results do not reveal the existence of any significant correlations between the mechanical properties of the rocks, which the aggregates are made of, and fracture energy and compressive strength of concrete'. In my opinion, it would require better-structured research to tease out the influence of the properties of the parent rock (such as hardness, density, mineralogy, crystal structure) upon crack development in concrete; such an influence cannot be absent, and will be considered later in this section where experimental data are available.

Types of cracks

The cracks considered in this section are those wider than about 0.1 mm, that is, those visible to the naked eye. Narrower cracks are referred to as microcracks. Microcracking consumes energy and therefore toughens the concrete and causes non-linear fracture behaviour [3].

Slate and Hover [4] describe microcracks as microscopic brittle failures, and their development under a progressively increasing load results in a pseudo-ductile behaviour, which produces the well-known shape of a 'bending over' stress–strain curve. This development results in constantly changing load paths and a decrease in the number of available load paths, each of which becomes subjected to an ever-increasing load [4]. These microcracks are of no direct relevance to the question posed in the title of the present section because they do not occur within the aggregate particles.

The stress–strain curve referred to above is that under a nominally uniaxial compression, which is, from the structural standpoint, the main mode of exploitation of concrete, tensile loads being generally carried by steel reinforcement. Yet the predominant failure mode is tension or tension-shear: this the transverse tensile strain resulting from the longitudinal compressive stress [4]. This strain results in a longitudinal crack, and this is why tensile strain capacity of concrete is of great importance [5].

This brings us to a consideration of the broad location of the cracks: on the exterior of the concrete element, visible for all to see, or in the interior of the concrete. It is the former cracks that we observe when inspecting concrete surfaces, but it is the latter cracks that are significant to structural failure. In a general discussion, we do not distinguish between the two loci of cracks but, as pointed out by Slate and Hover [4], the free surface of concrete has an entirely different stress field than the interior. For the purpose of this section, it is the crack paths in the interior that are of interest; the cast exterior generally does not show any coarse aggregate particles.

This section is concerned with the propagation and development of cracks and not with failure of a concrete member. However, failure is preceded by cracking, and there is no universal or all inclusive definition of failure. Zaitsev expresses the view that 'failure really means a gradual degradation of the structure' [6]. He also says: 'In a composite structure, failure is usually not caused by the extension of one crucial crack. In fact, well below the ultimate load, cracks begin to propagate and are arrested again by high-strength aggregate particles' [6]. This statement is clearly predicated on aggregate particles being the strongest element in the composite material, that is, concrete. In the case of lightweight aggregate concrete, Zaitsev [6] shows crack patterns of cracks originating from pores and running through the aggregate particles, but these are computer-generated and not confirmed by experiment. It is the lack of direct experimental evidence that has bedevilled my literature search. Furthermore, Zaitsev [6] does not comment on the relevance of the strength of lightweight aggregate particles or generalize the situation.

Directly relevant literature

Following the usual approach, I searched the literature for papers on crack paths, cracking of aggregate, fracture of concrete, and even on strength of aggregate particles. I found no direct answer to the question in the title of the present section.

I should add that I am concerned only with experiment-based papers or with analyses confirmed by experiments. This means that I have not studied extensively purely energy-based analyses or finite element analyses unrelated to the *actual* mechanical and physical properties and not confirmed by experiment; nor have I considered the solely scientific understanding of crack development.

The above exclusions do not mean that I ignored papers in those categories. Indeed, I found useful pointers in a number of papers, and I shall discuss them in the present section. Nevertheless, my aim is to obtain a practical review of a practical question as a macro-view of crack development.

One of the few papers explicitly discussing the question of cracks passing through coarse aggregate particles, is that by Remmel [7]. He tested high-strength concrete (122 MPa at 21 days) and found the cracks to pass through coarse aggregate particles, and ascribed this behaviour to 'a more equal strength of the aggregate and the matrix and a better bond between them' [7]. He also observed that 'the crack surface was relatively smooth' [7]. I interpret that to mean that the surface was more plane than a crack surface that meanders around coarse aggregate particles.

The nearest published recognition of the broad factors influencing the crack path was in a paper by El-Sayed *et al.* [8], who wrote: 'Depending on the strength of the aggregate and of its interface with the matrix, the crack propagation goes through the particles or around them, thus resulting in a different crack roughness.' It is not clear whether roughness refers to the planeness of the crack surface.

Wu *et al.* [9] recognized that, in high-strength concrete, 'cracks may extend through the aggregate, which makes use of the full strength potential of the coarse aggregate particles. Thus the fracture energy of the aggregate influences the fracture energy of concrete' [9]. This argument accords with my view that cracks may pass through aggregate particles when the strengths of the aggregate and of the matrix are similar.

General studies on cracking

The early and seminal work on cracking was the doctoral dissertation of Maso, published in 1969 [10]. He commented on failure paths in the interface zone of aggregate–paste and considered three explanations of the failure at the interface: (i) the hydrated cement paste in the interface zone is weaker than the bulk paste; (ii) there are discontinuities in the contact between the paste and the substrate aggregate; and (iii) the local stresses in the interface zone are higher than elsewhere. According to Maso [10], these three explanations are not mutually exclusive.

It is now known that the interface zone has a different microstructure from the bulk of the hardened cement paste; the interface zone is the locus of early microcracking, known as bond microcracking [11].

The first explanation of Maso, that is, the weakness of the interface zone, was studied in depth (no pun intended) and confirmed in 1993 by Larbi [12]; the difference in the compound composition of the paste in the interface zone, the orientation of the crystals of $Ca(OH)_2$, and the lower strength of this paste are now universally accepted. Parenthetically, it is worth remembering that the interface zone exists at the surface of all aggregate particles, including the fine particles, so that 30 to 50% of the total volume of the hydrated cement paste is of the interface quality [12]. However, for the purpose of considering the crack paths, only the interface zone around the coarse particles is of interest. Indeed, the number of fine particles of aggregate is so large that even a plane crack passes through the interface zone around fine particles.

The above observation gives rise to the suggestion that even a crack path remote from coarse aggregate particles follows the interface zone. From this, it is easy to speculate that *all* crack development is governed by the strength of the paste in the interface zone. This, however, is not of primary relevance to the topic of the present section.

The second explanation of failure at the interface, offered by Maso [10], is the presence of discontinuities there. These may be the consequence of air voids on the surface of large aggregate particles or bleed voids. Both types of voids are the consequence of the treatment of concrete during compaction (consolidation) and are not intrinsic properties of the material.

Difference in moduli of elasticity

The third explanation given by Maso [10], namely the elevated local stresses, is related to the difference between the values of the modulus of elasticity of the hydrated cement paste and that of the aggregate particle. Buyukozturk and Hearing [13] also commented on the relevance of the relative value of the modulus of elasticity of the aggregate and the matrix. This topic was studied by Neville [14], who found that this difference has a considerable influence on the bond between cement paste: the smaller the difference the smaller the local stress concentration.

It follows that failure of concrete due to excessive bond stresses is predominant in ordinary concrete, in which the modulus of elasticity of aggregate is considerably higher than that of the hydrated cement paste. Conversely, the modulus of elasticity of lightweight aggregate is generally not significantly different from the modulus of the cement paste. Likewise, in the case of very high strength concrete, in which the w/c of the cement paste is very low, the modulus of elasticity of the hydrated cement paste is only a little lower than the modulus of the aggregate [14]. In those situations, there is a strong likelihood of monolithic behaviour in concrete, coupled with a more linear stress–strain relation and increased brittleness [14].

The last situation is of especial interest in compression testing because the failure of the specimen can be very sudden and explosive with fragments of the shattering specimen flying in all directions.

The occurrence of explosive failure reminds me of my laboratory teaching days: testing concrete made with high-alumina cement (which develops a very high strength) was dangerous until we installed a screen around the specimen being tested.

Bond failure

Maso [10] studied the bond between the hydrated cement paste and aggregate, subjected to tension and to compression. What is of interest here is the change between the ages of 7 and 28 days. He found no difference in the products of hydration at the two ages, but there was an increase in the extent of the actual contact surfaces between the aggregate and the paste. He also looked at the influence on the bond strength of the mineralogical nature of the aggregate. This was observed also by Aïtcin and Mehta [15]. I shall return to the influence of the nature of the aggregate on failure under the heading of 'Influence of aggregate'.

Bond failure, apparent as a crack going around a particle of coarse aggregate, needs some elaboration. Larbi [12] refers to two opinions: according to one, cracks develop *at* the surface of the aggregate; according to the other one, the cracks develop *within* the interface zone, at a distance of a few micrometres from the surface of the aggregate.

The two possible areas of failure arise from the fact that there are two distinct parts of the interface zone. One area is the so-called duplex film, consisting of a layer of oriented crystalline $Ca(OH)_2$ adjacent to the aggregate surface and an adjacent layer of calcium silicate hydrate, each layer about $0.5\,\mu m$ thick. The second area is the outer part of the interface zone. Failure can occur in either of these two areas. For the purpose of this section, the distinction is of no importance.

Continuing hydration densifies the second area, thus improving its strength and resistance to cracking [16]; the duplex film does not change much. However, insofar as the subject of this section is concerned, namely the influence of age on cracking around the aggregate particles, the process of hydration continues over weeks and months, rather than years and decades. The topic of the increase in strength of concrete with age will be considered under the heading 'Effect of age on strength'.

A difficulty in assessing the bond strength is that it cannot be predicted from the aggregate surface roughness alone; the strength and structure of the parent rock are also factors influencing the bond strength [17], but no quantitative approach has as yet been developed.

What is the weakest component?

The influence of the actual strength of the aggregate particles on the strength of mortar is not clearly understood. Larbi [12] found that, at a given age and w/c, neat cement paste (made with Portland cement) has a higher compressive strength than mortar containing a similar paste. He pointed out that the particles of fine aggregate are 'considerably stronger than the cement pastes' [12]. I see no contradiction here because the particles of fine aggregate do not directly influence the strength of the composite, that is, mortar. What is critical is that the particles are stronger than the paste: how much stronger is not of importance. This is not the last word on the subject, but the topic of strength of hydrated cement paste *versus* the strength of mortar or concrete is outside the scope of this section.

Guinea et al. [17] considered the strength of the interface relative to the strength of the matrix and relative to the strength of the aggregate to be the critical factor determining whether or not the crack passes through the aggregate particles. The locus of the crack affects its roughness and influences the energy consumption, thus affecting the toughness of concrete [17]. They found that improving bond by the inclusion of silica fume in the mix resulted in a higher proportion of cracked aggregate particles [17].

However, Tasong et al. [18] pointed out that the interface zone is not necessarily the 'weakest link' in concrete. On the other hand Giaccio and Zerbino [19] go so far as to say that 'the interfaces are the weakest link in concrete'.

Despite the fact that, as early as 1969, Maso [10] clearly showed the important relevance of the transition zone to the crack path; 30 years later Yang and Huang [20] still discussed 'a model to predict compressive strength of cement based composite by considering concrete as a two phase material' [20]. Worse still, they quoted, with approval, papers that demonstrated that the strength of concrete is controlled by the weakest component' [20]. One of the components that they considered was coarse aggregate. Now, this 'weakest-link in a chain' approach cannot be valid, as demonstrated by the behaviour of cellular concrete, known also as gas concrete, aerated concrete, or foamed concrete. Strictly speaking, this cement-based composite material is not a concrete at all because no aggregate is present.

What is relevant to my argument is that the air voids are a component of 'concrete'. They are *the* weakest component, they have a zero strength, and yet the 'concrete' has a strength of several MPa, usually

2 to 10 MPa [5]. Admittedly, this is a very low strength but by no means the strength of the 'weakest component'.

It is not within the scope of this section to discuss the factors controlling the strength of cellular concrete. Nevertheless, it may be worthwhile to mention Hoff's suggestion that the strength can be expressed as a function of the void content taken as the sum of the induced voids and the volume of evaporable water [21]. Thus, the factor controlling the strength is the total volume of voids, and this, I believe, applies to all concretes. No-fines concrete is another example of concrete containing large air voids as the weakest component, but strengths up to about 16 MPa are readily obtained [5].

Influence of aggregate

Giaccio and Zerbino [19] recognized that coarse aggregate particles arrest crack growth but 'some particles are fractured. This mechanism depends greatly on the characteristics of the aggregate, especially surface texture and shape, and on the strength differences between aggregates and matrix' [19]. This last parameter is rarely mentioned, but it is crucial to the question of influence of the age of concrete upon crack paths.

Giaccio and Zerbino [19] tested several types of aggregate, both river gravel and crushed rock, and granite as well as quartz; these various aggregates were used in high-strength concrete ($w/c = 0.3$) and in ordinary concrete ($w/c = 0.5$). While generalizations should not be made from one set of tests with respect to specific rock types, their conclusions are of importance. First, the influence of the type of coarse aggregate is greater at high strengths of concrete. Second, when the matrix strength is close to the rock strength, the probability of cracks passing through aggregate particles increases. Third, in high-strength concrete, the interface strength has a large influence upon the failure behaviour of concrete; this strength is due to bond and mechanical interlocking [19]. I take the last factor to be the consequence of surface texture of the aggregate. I also note that the shape of the coarse aggregate particles varied between the various aggregate types, and this shape is relevant to the interpretation of the test results.

De Larrard and Belloc [22] found that for ordinary concrete (with aggregates from a parent rock), the strength of the concrete is governed by the strength of the matrix, being a modified value of the latter; two parent rocks were used, one with a strength of 285 MPa, the

191

other 250 MPa [22]. We can add that, almost 50 years ago, Neville [23] quoted Kuczynski to the effect that, at high values of w/c, the strength of aggregate or its shape has no influence on the strength of concrete.

According to de Larrard and Belloc [22], bond between the aggregate and the matrix influences the strength of concrete, calcareous aggregate being particularly good in this respect. For high-strength concrete, there is a ceiling due to the intrinsic strength of the parent rock; however, the actual strength of the concrete is lower than the strength of the rock itself. The paper by de Larrard and Belloc [22] does not give any information about the crack paths, but we can infer that, when the strength of the aggregate is the controlling factor, at least some of the cracks pass through aggregate particles.

There is a negative consequence of cracks passing through limestone aggregate. In the case of high-strength concrete (cube strength of 100 MPa) used in deep beams without shear reinforcement, the crack surfaces are too smooth for shear transfer. In consequence, the shear strength assumed in design should be taken as that corresponding to concrete with a cube strength of 60 MPa. This topic, of importance in structural design, is still to be studied fully by the Concrete Society.

Buyukozturk and Hearing [13] tested notched beams with an inclusion above the notch of one or two cylindrical particles fashioned from rock. When the inclusion was of high-strength granite (123 MPa), the crack passed through the interface zone; with low-strength limestone (57 MPa) the aggregate inclusions were fractured. In the specimens that developed interface cracks, the failure load increased with an increase in the strength of mortar and with an increase in the roughness of the surface of the aggregate. The strength of mortar also had a positive influence on the load-carrying capacity of specimens in which the cracks passed through the aggregate particles.

More generally, Buyukozturk and Hearing [13] observed that in high-strength concrete, cracks propagate through the aggregate particles, which thus cannot act as crack arresters. They used manufactured 'aggregate' particles, but with naturally occurring aggregate it is not possible to determine the strength of the parent rock.

That the energy required to fracture coarse aggregate particles, as compared with a crack extending around the particles, is larger was confirmed by El-Sayed *et al.* [8]. The effect is larger with crushed aggregate than with rounded particles because, in the former, the ratio of broken surface to the debonded surface is larger [8]. In other words, more cracks travel through the crushed aggregate than when the aggregate is rounded.

A possible starting experiment to determine the unconfined compressive strength of a particle of coarse aggregate, say nominally 60 mm, would be to place it in a mould that is just large enough to hold it. The vacant space in the mould could be filled with a material of high strength but with a low modulus of elasticity. The failing axial load could easily be determined, but converting this value into a stress (in MPa) would require careful consideration.

Buyukozturk and Hearing [13] studied the crack paths in concrete composites and established the influence of the relative magnitude of the fracture toughness of the interface zone in relation to the values of fracture toughness of the aggregate particles and the matrix [13]. They also commented on the relevance of cracks travelling through coarse aggregate particles, which become fractured.

Wu *et al.* [9] confirmed that high-strength concrete, where both the strength of the matrix and the bond strength are similar to the strength of coarse aggregate, is rather brittle, which is undesirable.

Aggregate shape

When the properties of the parent rock are not a factor, that is, when both rounded particles and crushed particles originate from the same parent rock, the influence of the aggregate shape and surface characteristics upon the proportion of cracked particles can be determined. Such a determination was carried out by Guinea *et al.* [17], who found a very much higher occurrence of cracks passing through crushed aggregate. This is not surprising but experimental support is valuable. We should note, however, that this effect is limited to concretes with a splitting tensile strength of less than about 3 to 4 MPa [17]. At higher strength, the proportion of cracked particles of large aggregate was found to be less than 0.2 [17]. More systematic studies of the consequences on cracking of shape and texture are not available.

Tests on wedge splitting of specimens of concrete used in dam construction, with a maximum aggregate size of 38 mm, showed that, with rounded aggregate, debonding was the main mode of failure; with subangular particles, aggregate fracture was prevalent [24]. In the latter case, more energy was required than in the case of debonding [24]. These findings are unremarkable, but are reported in view of the paucity of experimental results.

Aggregate size

It is well known that a larger maximum aggregate size reduces the water requirement of the mix. However, the use of such an aggregate is counterproductive when it is desired to produce high-strength concrete. This is due to the development of more bond microcracking when the maximum aggregate size exceeds about 10 mm [11]. Larger size of aggregate may also lead to more bleeding voids and, therefore, unbonded surfaces.

Hiraiwa et al. [25], who used a casting method that did not result in bleed water collecting underneath the large aggregate particles, found no adverse effect of using a larger maximum aggregate size.

Nevertheless, bleeding occurs quite often, and more generally, a smaller maximum size offers also a larger total surface area and therefore a lower bond stress, so that bond failure does not occur [11]. Consequently, cracks pass through the aggregate, both under compressive and under tensile loading [7].

Lightweight aggregate

Experiments on concrete made with lightweight aggregate are revealing because such aggregate has a low strength, often lower than the strength of the matrix. lightweight aggregate usually also has a high surface roughness, which leads to good bond with the matrix. Such experiments were performed by Husem [26]. He used dasitic tuff as lightweight aggregate so that he could determine its compressive strength (39 MPa) on the parent rock; the strength of the concrete made with that aggregate (at a $w/c = 0.50$) was 18 MPa. Unfortunately, he did not report any observations on the development of cracks in his specimens, and his conclusions about the additive nature of the strength of the matrix, the strength of the aggregate, and the bond between the two are, in my opinion, highly suspect.

I have referred earlier to a paper by Yang and Huang [20] because it is one of the very few papers that consider the strength of lightweight aggregate, but they send confusing signals: as already mentioned, they consider the strength of the concrete to be 'controlled by the weakest component,' but they also say that 'both matrix strength and composite strength are much higher than the lightweight aggregate strength'. What is relevant to the present section is that Yang and Huang [20] observed cracks passing through coarse aggregate particles, and this supports the argument that the crack path is sensitive to the relative strengths of the matrix, the coarse aggregate, and the interface between the two.

Effect of age on strength

Modern Portland cements, with a high C_3S/C_2S ratio and a high fineness, develop strength rapidly so that there is not much increase in strength beyond the age of 28 days. On the other hand, when other cementitious materials, such as fly ash and ground granulated blast-furnace slag, are included in the mix, there may be a significant increase in strength beyond 28 days. In such concretes, the increase in strength beyond the age of 28 days varies considerably, often being 50% more than with Portland cement only.

Because there has been a suggestion of the possibility of dating of cracks that are 10, 20, or even 30 years old, it is useful to consider the increase in strength with age of concrete made with 'old' Portland cements. In design, the gain in strength beyond the age of 28 days is usually considered a 'free' contribution to safety, but the British Code of Practice for Reinforced Concrete in Buildings, CP114 [27], published in 1957, allowed the gain in strength to be taken into account in the design of structures that would not be subjected to full load until a later age. Table 5.1.1 shows the relevant values.

The practical significance of Table 5.1.1 is that, if the strength of the concrete at the age of one month, f'_{28}, was such that cracks would develop within the matrix or at the interface, cracks would pass through the coarse aggregate particles at the age of 12 months or greater only if the strength of the parent rock, f'_r, were not greater than the strength of the matrix at 12 months. In other words, cracking of the aggregate would occur only if the strength of concrete at 12 months, or later, f'_{365}, is such that $f'_{365} > f'_r$ and $f'_r > f'_{28}$. These values are, of course, broad estimates.

Table 5.1.1. Age factors for compressive strength according to CP 114 (1957) [27]

Age: months	Age factor
1	1.00
2	1.10
3	1.16
6	1.20
12	1.24

Tests examining explicitly the change in the concrete fracture parameters with age were performed by Zollinger *et al.* [3]. They found that young concrete is more brittle than mature concrete, but the greatest age investigated was only 28 days. Their review of other tests, of somewhat greater duration, reports that the critical stress intensity increases slightly between the ages of 28 and 90 days, but only when river gravel aggregate is used; with crushed limestone, there was no increase [3]. Also, between the ages of three and 28 days, the increase in the critical stress intensity was about twice as large (54%) with river gravel as with crushed limestone (23%) [3]. However, the value of the critical stress intensity depends also on the specimen geometry and size, so that this parameter is not a material constant for concrete [3].

Sarkar and Aïtcin [28] recognized explicitly the influence of age upon the locus of failure. However, they considered only very early ages. For example, they say that at the age of one day, whatever the water–cement ratio, cracks occur within the hardened cement paste and extend around the aggregate particles. They ascribe this to the low degree of hydration of Portland cement and the absence of the formation of products of reactions of other cementitious materials [28]. These statements are irrefutable, but the present section is concerned with the effects of age of concrete in service, which runs in years.

Zollinger *et al.* [3] extended their tests to derive fracture parameters that are material constants and calculated the fracture toughness (that is, the resistance to crack propagation) in an infinite specimen and the effective crack length; this is the fracture process zone, or a non-linear region in front of the crack tip in which toughening of the concrete takes place. They found that both the fracture toughness and the effective crack length increased between the ages of 12 hours and 28 days [3]. The fracture toughness increased faster than the compressive strength [3].

We can add that, according to Xi and Amparano, the size of the fracture process zone decreases with an increase in the relative volume of aggregate in the concrete [29].

I would like to mention the difference in the behaviour of concrete made with limestone aggregate and with gravel, referred to above [3]. At the age of one day, in the former, the cracks passed through all the coarse aggregate particles in the crack path; on the other hand, with gravel aggregate, very few particles were cracked.

A corollary of all the above is that, with strong aggregate and weak bond, such as gravel, which has a smooth surface, cracks did not pass through the aggregate. Also, the increase in critical stress intensity with age was large because the hardened cement paste became stronger. On the other hand, with limestone aggregate, the toughness of the aggregate continued to be relatively low. It follows that the strength of the coarse aggregate (which cannot readily be determined) and its bond characteristics are relevant to the influence of the increase in the strength of the matrix upon the crack path: when the strength of the aggregate is high and bond is poor, an increase in the strength of the hardened cement paste should not be expected to lead to cracking of the coarse aggregate particles. Hence, the presence of cracks around the aggregate particles cannot be interpreted as recent cracking; nor can it say anything about the age of cracks.

Conclusions

This section is not a conventional state-of-art review because there are hardly any published papers addressing the question in the title of this section. What prompted me to write it is the proposition made recently that the crack path is governed by the age of the concrete at which the crack developed. A literature search of papers with varying objectives and widely differing parameters has confirmed that, for ages of practical interest, there is no simple relation between the age at which a crack has developed and the nature of the crack: within the matrix, at the aggregate-matrix interface, or through coarse aggregate particles. The locus of the crack is governed by the relative strengths of the three loci and by the fracture energy required for the crack development.

Admittedly, between the ages of, say, one day and a week or two, the development of the crack is likely to be influenced by the increase in the strength of the matrix and in the bond strength, but this is of no practical interest. Beyond the age of about 28 days, the locus of cracking is not influenced by the increase in the strength of the matrix because the balance of strength between the aggregate particles themselves and the matrix, or between the aggregate particles and the bond between the two, does not change, the increase in the strength of the matrix being small.

The second prompt for this section was that, as recently as 2004, there was proposed an amendment to the Concrete Society guidance for high-strength concrete made with limestone aggregate, recognizing the fact 'that aggregate fracturing at cracks had produced crack surfaces too smooth for the shear transfer implicit in design recommendations' [1]. This observation shows that the strength of the aggregate is only one of the factors involved; the type of the rock is also relevant.

The locus of cracks is influenced also by the characteristics of the coarse aggregate particles, specifically, their shape and surface texture; these may be influenced by size in consequence of differing processes of comminution of the rock.

We can see, therefore, that the path taken by cracks in concrete is not just of academic interest but has a structural significance as well. However, a quantification of the factors influencing the interlock is not available.

These conclusions, limited as they are, represent fully the current situation, and the present section is probably the first complete review of publications on *experimental* work. Admittedly, analytical work on cracking has been done, but it has not been verified or confirmed by experimental data. The credence of such analyses is, therefore, limited.

In fact, there is no coherent experimental evidence available to answer the question in the title of the section: which way do concrete cracks run? We have to content ourselves by saying, 'see how they run', which is precisely what Thomas Ravenscroft wrote in 1609 in *Deuteromelia*, now a well-known nursery rhyme: 'Three blind mice (or cracks?), see how they run!'

If we were able to include the consideration of the crack path in selection of the mix ingredients, we would be able to ensure a more ductile failure, which is structurally desirable. This objective cannot be achieved on the basis of limited-scope experiments using a very small number of combinations of aggregate type, shape, and surface. Unique and usable relationships are unlikely to be found by such small experiments (except for a local situation), and there is a danger of findings of spurious relationships, such as the crack path and the age of concrete, entering the literature. Large projects are difficult to organize and finance because of the considerable range of the size of aggregate particles and of the associated surface characteristics, which depend not only on the parent rock but also on the method of comminution of the particles. Moreover, for many real-life aggregates, there is no parent rock that can be tested.

Nevertheless, a major coherent research project is needed to enable us to improve our knowledge of crack paths. Concrete is a wonderful construction material because it does not disintegrate suddenly or explosively and, with improved knowledge of crack paths, we should be able to design economical mixes with balanced properties of strength and durability better. And we shall then know which way the concrete cracks run.

References

1. Amendment to the Concrete Society Technical Report 49, Design guidance for high strength concrete, *Concrete*, 38, 7, 2004, p. 4.

2. Hassanzadeh, M., The influence of the type of coarse aggregates on the fracture mechanical properties of high-strength concrete, *Fracture Mechanics of Concrete Structures*, Proceedings FRAMCOS-3, Aedificatio, Freiburg Germany, Vol. 1, 1998, pp. 161–170.

3. Zollinger, D.G., Tang, T. and Yoo, R.H., Fracture toughness of concrete at early ages, *ACI Materials Journal*, 90, 5, 1993, pp. 463–471.

4. Slate, F.O. and Hover, K.C., Microcracking in concrete, *Engineering Application of Fracture Mechanics*, 3, 1984, pp. 137–159.

5. Neville, A.M., *Properties of Concrete*, Fourth Edition, Wiley, New York, 1996, 12th impression, with standards updated to 2002, Pearson Education, 2005.

6. Zaitsev, Y., Crack propagation in a composite material, in: *Fracture Mechanics of Concrete*, Wittman, F.T. (ed.) Elsevier, Amsterdam, 1983, pp. 251–299.

7. Remmel, G., Study on tensile fracture behaviour by means of bending tests on high strength concrete (HSC), *Darmstadt Concrete*, 5, 1990, pp. 155–162.

8. El-Sayed, K.M., Guinea, G.V., Rocco, C.G., *et al.*, Influence of aggregate shape on the fracture behavior of concrete, *Fracture Mechanics of Concrete Structures*, Proceedings FRAMCOS-3 Aedificatio, Freiburg Germany, vol. 1, 1998, pp. 171–180.

9. Wu, K.-R., Chen, B., Yao, W. and Zhang, D., Effect of coarse aggregate type on mechanical properties of high-performance concrete, *Cement and Concrete Research*, 31, 10, 2001, pp. 1421–1425.

10. Maso, J.-C., La nature minéralogique des agrégats: facteur essentiel de la résistance des bétons à la rupture et à l'action du gel, *Ciments et Bétons*, 647–648, 1969, pp. 247–276.

11. Neville, A., *Neville on Concrete*, ACI, Farmington Hills, MI, 2003.

12. Larbi, J.A., Microstructure of the interfacial zone around aggregate particles in concrete, *Heron*, 38, 1, 1993, pp. 1–69.

13. Buyukozturk, O. and Hearing, B., Crack propagation in concrete composites influenced by interface fracture parameters, *International Journal of Solids and Structures*, 35, 31–32, 1998, pp. 4055–4066.

14. Neville, A.M., Aggregate bond and modulus of elasticity, *ACI Materials Journal*, 94, 6, 1997, pp. 71–74.

15. Aïtcin, P.-C. and Mehta, P.K., Effect of coarse-aggregate characteristics on mechanical properties of high-strength concrete, *ACI Materials Journal*, 87, 2, 1990, pp. 103–107.

16. Liao, K.-Y., Chang, P.-K., Peng, Y.-N. and Yang, C.-C., A study on characteristics of interfacial transition zone in concrete, *Cement and Concrete Research*, 34, 6, 2004, pp. 977–989.

17. Guinea, G.V., El-Sayed, K., Rocco, C.G., *et al.*, The effect of the bond between the matrix and the aggregates on the cracking mechanism and fracture parameters of concrete, *Cement and Concrete Research*, 32, 12, 2002, pp. 1961–1970.

18. Tasong, W.A., Lynsdale, J. and Cripps, J.C. Aggregate-cement paste interface. II Influence of aggregate physical properties, *Cement and Concrete Research*, 28, 10, 1998, pp. 1453–1465.

19. Giaccio, G. and Zerbino, R., Failure mechanism of concrete: combined effects of coarse aggregates and strength level, *Advances in Cement Based Materials*, 7, 1988, pp. 41–48.

20. Yang, C.-C. and Huang, R., Approximate strength of lightweight aggregate using micromechanics method, *Advances in Cement Based Materials*, 7, 1998, pp. 133–138.

21. Hoff, G.C., Porosity – strength considerations for cellular concrete, *Cement and Concrete Research*, 2, 1, 1972, pp. 91–100.

22. de Larrard, F. and Belloc, A., L'influence du granulat sur la résistance à la compression des bétons, *Bulletin Laboratoires des Ponts et Chaussées*, 218, 1999, pp. 41–52.

23. Neville, A.M., *Properties of Concrete*, First Edition, Pitman, London, 1963.

24. Saouma, V.E., Broz, J.J., Brühwiler, E. and Boggs, H.L., Effect of aggregate and specimen size on fracture properties of dam concrete, *ASCE Journal Materials Civil Engineering*, 3, 3, 1991, pp. 204–218.

25. Hiraiwa, T., Tanigawa, Y., Mori, H. and Nanbu, Y., Study on effects of size and arrangement of coarse aggregate on compressive fracture behaviour of concrete, *Japan Concrete Institute*, 20, 1998, pp. 67–74.

26. Husem, M., The effects of bond strengths between lightweight and ordinary aggregate mortar, aggregate-cement paste on the mechanical properties of concrete, *Materials Science in Engineering*, A363, 2003, pp. 152–158.

27. British Code of Practice CP 114, *The Structural Use of Reinforced Concrete in Buildings*, 1957.

28. Sarkar, S.L. and Aïtcin, P.-C., The importance of petrological, petrographical and mineralogical characteristics of aggregates in very high strength concrete, *Petrography Applied to Concrete and Concrete Aggregates*, ASTM STP 1061, Philadelphia, PA, 1990, pp. 129–144.

29. Xi, Y. and Amparano, F.E., Effect of aggregates on fracture process zone of concrete, *Engineering Mechanics*, 11th Conference, ASCE, Vol. 2, 1996, pp. 1185–1188.

ADDENDUM

In Section 5.1, I referred to the negative effect of the smooth surface of the crack passing through limestone aggregate particles on the shear transfer in a diagonal crack in a beam. A recent paper by P.E. Regan *et al.* [1] reports the results of a systematic study of beams made of concrete containing limestone aggregate and concludes the 'the shear strengths of members without shear reinforcement are often below characteristic resistances calculated according to [the code] EC2 and other recommendations. A considerable proportion of the experimental strengths can be below design resistances. The deficits of resistance are greatest where high concrete strengths are combined with relatively large effective depths'. This situation is particularly grave because, unlike flexural failure involving yield of reinforcement which is ductile, shear failure is sudden, without warning, and often catastrophic.

Regan *et al.* [1] observed that 'the crack surfaces in the beams with limestone aggregate are extremely smooth'.

It is worth noting that the findings of Regan *et al.* have been confirmed in tests in the USA and Canada. All this applies to concretes with a compressive strength limited to 50 MPa measured on cubes, or to 60 MPa on cylinders.

Reference
1. Regan, P.E., Kennedy-Reid, I.L., Pullen, A.D. and Smith, D.A., The influence of aggregate type on the shear resistance of reinforced concrete, *The Structural Engineer*, 6 Dec., 2005, pp. 27–32.

5.2 SOME ASPECTS OF SUSTAINABILITY

The concept of sustainability began to loom large following the Report of the 1987 World Commission on Environment and Development, specifically the so-called Brundtland Report. The crucial words are: 'Humanity has the ability to make development sustainable – to ensure that it meets the needs of the present without compromising the ability of future generations to meet their own needs.'

Prior to January 2006, I had not published any papers on sustainability but, finally, I felt I could not ignore that topic in this book because it has become the flavour of the decade, and we are told that we must all be supporters of sustainability. I am unable to add to the rhetoric, if not clamour but, to redress the balance, a few cool comments may be useful.

Parenthetically, I could add that the grip of fear of running out of primary materials is not new. Soon after World War II, the so-called Club of Rome presented arguments about the dangers of running out of primary materials, such as coal, oil, iron ore, and some other metal ores. In most cases, the argument was based on known reserves, existing methods of extraction, and a continuing increase in population with the same consumption rate per capita. None of these proved to be a safe assumption. For example, not only new reserves of metal ores have been discovered, but also the methods of extraction have been improved so that old mine tailings are now being re-worked to extract the metal. Moreover, in many applications, such as motorcars, steel has been replaced by plastics and fibres. Likewise, plastics are now used in some furniture, underground storage tanks, drains, pipes, and fabrics. The application of composites to aircraft wing technology is being developed. So the old fears were largely unfounded, and the Club of Rome died a quiet death.

Only the population growth continues, especially in Asia, but this is too sensitive a topic in religious and moral terms for me to broach. I shall limit myself to saying that these mega-nations will demand, and rightly so, provisions of modern life not only for a larger number of people but also at the standard of the currently rich countries.

Against this backdrop, our local application of sustainability will prove puny, but this does not mean that we need not do our best here at home.

To start with generalities, of course, we want the world to be run in a sustainable manner, by which I mean not being wasteful of materials or energy and not degrading the environment. As Broers [1] put it, 'the sustainable development concept requires of all of us – as engineers and citizens – to consider much more widely than before the impact of our own lives and of the infrastructure and products we produce ...'. He then continues: 'engineers need to integrate consideration of whole-life environmental and social impacts – positive as well as negative – with the mainstream and commercial aspects of their work. Wise use of natural resources, minimum adverse impact and maximum positive impact on people and the environment are the targets.'

Sustainability of concrete construction

This book is about concrete, and I should therefore limit myself to sustainability of concrete and concrete construction. Strictly speaking, the term construction subsumes concrete as a material because our interest in the material *per se* is limited. It follows that, wherever possible, we should improve the production methods of concrete and of cementitious materials, and we should also design concrete structures so that their operation is conducive to minimizing energy consumption and that they are adequately durable. In my opinion, there is no such thing as inherently durable concrete. There is durable concrete for the purpose: it should perform its intended function during its intended life, but not necessarily any longer.

To promote sustainability, the approach has to be such that all aspects of concrete structures are taken into account. Taking energy as an example, we should consider not only the energy efficiency of the operation of a building but also the energy aspects of the materials used and of the process of construction. Alas, those who want to show themselves to be on the side of the angels of sustainability do so by a partial and selective approach and even by slinging the proverbial mud at competing types of construction. Such an approach does not enhance the image of concrete and, indeed, tarnishes it in the eyes of intelligent and informed readers.

Let me give an example of what I consider crude over-selling coupled with casting of wholesale aspersions on steel structures. Before I report some details of the statements in what I consider to be the offending article, I have to say that it was not published in a commercial journal, but in a highly reputable periodical, *Science in Parliament*, which is the Journal of the Parliamentary and Scientific Committee, whose objectives are: '(a) to inform the scientific and industrial communities of activities within Parliament of a scientific nature and of the progress of relevant legislation; (b) to keep Members of Parliament abreast of scientific affairs.' The journal probably does not have a wide circulation but it is highly influential, being the forum for views of scientists and engineers for consumption by parliamentarians. As a member of the Parliamentary and Scientific Committee, I am aware of the situation, and I deplore even more profoundly the crude approach of 'character assassination' of steel while allegedly promoting concrete.

Behaviour of concrete in fire

I shall refer to only a few aspects of the behaviour of concrete in fire, good fire resistance being essential for a given concrete structure to satisfy the desiderata of safety. This section is not an exposé on the behaviour of concrete in fire.

The title of the article that I intend to criticize is 'Steel's fire performance under scrutiny' and the author is a chemist who is the Head of Performance at the Concrete Centre. Structural engineers know that what is crucial is not just the fire resistance of steel as a material as compared with the fire resistance of concrete. It is known that steel has to be fireproofed, but it should not be forgotten that the comparable structural material is not just concrete, but reinforced concrete. The reinforcement is obviously steel and, once sufficiently high temperatures have been reached, the concrete cover to reinforcement may spall, and the exposed steel then becomes vulnerable to fire because it is...steel. Concrete with an extremely low water–cement ratio, that is, very high-strength concrete, is particularly vulnerable in this respect.

The description of the consequences of fire on the Windsor Tower in Madrid by Anna Scothern [2] to the effect that 'Failure was limited to the perimeter steel frame whereas the internal concrete frame survived complete burn-out with no collapse' may well be true, but it is not an appropriate comparison of steel and concrete with respect to fire endurance.

We ought to rejoice that concrete as a material behaves well in fire because it is incombustible, does not emit toxic gases, and offers reasonable insulation against transmission of heat. But in structural design – and structures are the most important end use of concrete – we are concerned with fire endurance rating, which is the survival time of specific structural assemblies or components. Fire endurance rating is strongly affected by the properties of coarse aggregate used, by the thickness of cover to any embedded steel and, very importantly, by the structural system and design details, including restraint during heating in a fire.

But this is not all. Large structures, such as the World Trade Center in New York and the Windsor Tower in Madrid, have to be robust and to contain numerous redundancies to prevent progressive collapse, such as occurred at Ronan Point in England in 1968. Assessment of fire resistance of a structure, as against a single structural element, is a task for a structural engineer and requires careful and impartial approach. As someone who has spent the best part of half a century studying concrete, and establishing its properties and, thus promoting it, I find it a pity that all this endangered by an undocumented accusation of the inadequacies of steel, and thereby of superiority of concrete.

Concrete v. steel

With respect to fire resistance, concrete and steel are not simply interchangeable so as to optimize the fire resistance of a structure. Nor do they have the same mass for a given applied loading, so that the size of the foundations is likely to be affected by the choice of the material for the frame. Larger or more massive foundations mean more labour, more concrete and, therefore, more energy used in the construction, and possibly more carbon dioxide omissions in the production of the cement to be used in the concrete foundations.

Parenthetically, I would add that, in recent years, we have found it increasingly easier to make very high strength concrete, even 140 MPa or so. Alas, the design methods, and certainly design codes, have not kept up with the progress in concrete technology, so that we often do not derive full benefit from producing very high strength concrete.

More generally, design of concrete structures is more laborious and time consuming than steel design. As they say, time is money so that time represents cost, and this should not be ignored in any comparison of different types of structures. Nowadays, we should also say: energy is money. While we want to economize on energy and therefore on emissions of noxious gases, we cannot disregard the direct expenditure on personnel, computer time, office operation, etc.

Concrete and timber

I suppose that everyone who becomes more and more of a specialist in a chosen material lauds it enthusiastically. Nevertheless, we still ought to be careful not to lose sight of reality.

A recent example of a lack of care is a booklet entitled *Structural Design in American Hardwoods*, published in late 2005 by the American Hardwood Export Council. In the 'Foreword', Richard Harris says: 'There is no other construction material that comes anywhere near timber in its potential to produce environmental benefit. Because both in growth and in use timber is a carbon sink, the use of timber ... is always good environmentally...'. My concern is with the assertion that timber *in use* is a carbon sink. As I see it, while timber is growing, its leaves use carbon dioxide to build more wood: this process of photosynthesis has been known for a long time. Thus, growing timber is a carbon dioxide *sink*. After felling and processing into building material, the amount of fixed carbon dioxide remains constant, that is, timber becomes a carbon *store* while it remains in the structure.

I believe that Harris is aware of the distinction between the sink and the store but, in the booklet, enthusiasm went ahead, combining the benefits of re-forestation, and therefore creation of sinks, with the continuing store of carbon dioxide in the structure. Those of us who think of concrete as a material with sustainable characteristics need not overstate our case.

Recycling and concrete

Because sustainability has become everybody's goal, it is a lively topic. However, the approach is fragmented. Basically, we seek to economize energy in the production of new construction materials. Energy used to manufacture Portland cement includes, nowadays, a wide range of combustible materials, including car and truck tyres, but this does not significantly conserve fossil fuel, that is coal or natural gas, and does not greatly reduce the emission of carbon dioxide. But, of course, it is a move in the right direction.

So is recycling of aggregate and of concrete. However, we have to blend recycled material with virgin aggregate in order to have a workable mix. Recycling requires crushing and, therefore, transportation to the crusher, and back from it. Ignoring the energy involved in transportation as well as the more general cost of stockpiling, removal of steel, grading, and quality control of the recycled material does not present a true picture of the situation. Moreover, the shape and texture of aggregate consisting of crushed concrete are such that there may be a considerable influence on workability and, hence, appropriate measures have to be taken.

Precast concrete

The topic of recycling easily leads to consideration of precast concrete with respect to sustainability. In a precasting plant, there is less concrete waste, and such waste as there is can be easily recycled.

There is also economy in the *actual* amount of cement used per cubic metre of concrete because the difference between the design (minimum) strength and the actual mean strength is smaller consequent upon the lower variability of concrete produced in a fixed plant, as compared with site mixing. The energy saving is considerable because the main energy consumed in producing concrete is in the manufacture of cement: the energy required to manufacture the cement is about three-quarters of the total energy used in making the given concrete.

Another benefit of precasting lies in closer control of the concrete cover to reinforcement, so that excessive cover (to be on the safe side) is not necessary.

It may also be useful to extend the above argument by re-stating the fact that the outermost part of a concrete member is critical in resisting external chemical attack and, therefore, the quality of the finish is important with respect to durability. Precast concrete gives, in general, a superior finish because workmanship in a precasting plant is usually better than on a 'movable' site. In consequence, precast concrete generally has an enhanced durability (see Section 6.2).

Fly ash

Returning to the binders in concrete, the days of using pure unadulterated Portland cement have gone. Most mixes used nowadays contain fly ash or ground granulated blastfurnace slag, or both. (This is in addition to 5% of inert filler in so-called pure Portland cement.) These were waste materials from other processes but, with careful control or beneficiation, they are useful by-products. But 'careful control' does not come free and without knock-on effects. For example, to produce good-quality fly ash at the point of exit from combustion imposes restraints on the burning process.

Among the desiderata for high-quality fly ash, there is the need for a large proportion of vitreous spherical particles and also of particles smaller than $10\,\mu m$. These features are achieved by the use of very high burning temperatures in the coal-powder burning power-generating stations. However, the need to reduce the emission NO_x gases requires the use of lower peak burning temperatures with the consequence that minerals with a high melting point do not always fuse completely. With luck and ingenuity, a technology satisfying both the environmental requirement of limited NO_x emissions and the concrete user's need for small and spherical fly ash particles will be developed.

207

A further complication arises from the fact that an increasing number of power stations are mixing coal with biomass, wood in various forms, paper pulp, and vegetable fibre; European standards limit the mass of ash from these materials to 10% of the total ash. These changes will affect the use of fly ash in the concrete binder.

However, the biggest particular problem is the relatively low percentage of use of fly ash in the world. In many countries, including the USA, some power generating stations simply do not bother to produce fly ash suitable for use in concrete because this would require strict burning controls, including the use of coal from a single source. Progress is being made, but there is a long way to go before fly ash is not just a waste.

Use of tyres in a cement kiln

Tyres can be fed into the kiln as fuel in shredded form as a continuous process or intermittently through a special opening at the level of the burner, one or two tyres being inserted per revolution of the kiln. While the calorific value of tyres burnt in the kiln is high (if anything, higher than coal on a mass basis) there are costs associated with the process. In some countries, e.g. Quebec, Canada, there is a subsidy paid by the government (I believe $50 per tonne of tyres) so that from the standpoint of the cement manufacturer the tyre fuel represents a negative cost. Also, compared with coal, there is the benefit of reduced carbon dioxide emission. The residual matter from tyre combustion, which enters clinker, consists of sulfur, limestone filler and iron; these are, by and large, beneficial.

Thus, the use of tyres in the manufacture of cement is beneficial from the point of view of preserving natural resources (fossil fuel) and reducing carbon dioxide emissions. But in broad terms, this is just the proverbial drop in the ocean because cement kilns and tyre depots may be far apart. Traditionally, cement plants are located in the vicinity of exploitable raw materials for the manufacture of cement, and such a location may or may not be near the source of waste tyres. This, of course, is inevitable, but I am mentioning it because some enthusiasts claim very large benefits in terms of sustainability of concrete due to using tyres as kiln fuel. Even more outlandish claims are sometimes made.

For example, it is not correct to say: 'Within kilns, the steel content of tyres provides an essential source of iron, avoiding the need for shales and clays' [3]. Now, shales and clays are fed into the kiln to provide alumina and silica, which together represent typically 20 to 33% of the mass of Portland cement. If the entire fuel consisted of tyres, in a modern cement plant, only about 100 kg would be required to burn the raw materials for 1 tonne of cement. So shale or clay is still necessary.

This may be an appropriate place to mention that the production of one tonne of Portland cement requires about $1\frac{1}{2}$ tonnes of limestone and clay [4]. Another 'admission' that should be made is that the production of Portland cement generates about 1 tonne of carbon dioxide. A further admission is to say that, worldwide, the manufacture of Portland cement is reported to be about 8 to 10% of all man-generated carbon dioxide.

General

In my opinion, care is required in making sweeping statements, such as 'thermal mass of concrete offers the opportunity to reduce the energy requirements of buildings' [2]. The same applies, of course, to brick-work, and is the underlying principle of night-storage heaters. But what happens in practice depends on the periods of radiation reaching the concrete and of radiation *by* the concrete in relation to the indoor temperature required. I accept that advantage should be taken of the thermal properties of concrete, but architects are often lax in exploiting this.

Enthusiasm exceeding technical knowledge is also unhelpful. At a recent conference on sustainability at the Royal Academy of Engineering, a speaker urged us to design new buildings with a provision for a change of use so that, for example, a factory could be, at a later date, refitted to become a block of flats. Although refitting is often advantageous, provision during the initial construction is unrealistic because of the social requirements for accommodation: headroom, sanitation, fenestration, staircases, etc. Moreover, the capital cost of providing for a possible *future* use is generally unacceptable to the first owner. Indeed, the distinction between initial cost and later costs (including maintenance) is an important factor in design.

Sustainability is a multifaceted problem and we are as yet nowhere near an approach which takes all its aspects into account.

They say that charity begins at home. The American Concrete Institute has so far not produced a single guide on taking sustainability into account in concrete construction but in 2005 it decided to 'encourage the development of sustainable structures...'. I think that the UK is more advanced in this respect, and the Royal Academy of Engineering is conducting several in-depth studies, but no definitive views on sustainable concrete construction are available. We still have to await a cool wide-ranging appraisal. At present, the best overview available is that written by Holland [4] in 2002 who points out that in the USA 'there is no national activity geared toward specifying or certifying structures for their impact on the environment' [4].

References

1. Broers, Lord, Foreword, *Engineering for Sustainable Development: Guiding Principles*, The Royal Academy of Engineering, London, Sept., 2005, p. 3.
2. Scothern, A., Concrete is the best bet for sustainable development, *Science in Parliament*, Summer, 2004, pp. 16–17.
3. Hird, A.B, Griffiths, P.J. and Smith, R.A., Tyre waste and resource management: a mass balance approach, *Viridis Report VR2*, TRL Ltd, 2002, www.trl.co.uk.
4. Holland, T.C., Sustainability of the concrete industry: what should be ACI's role?, *Concrete International*, 24, 7, 2002, pp. 35–40.

5.3 REQUIREMENTS FOR RESIDENTIAL SLABS ON GRADE: APPROACH OF ACI AND OF UNIFORM AND INTERNATIONAL CODES

During the last ten years, I have been involved in the assessment of the adequacy of slabs on grade for residential single-family houses. One might think that, from the structural viewpoint, nothing could be simpler, and yet there has been much litigation; for my part, I have become progressively more bemused. Some expert witnesses have expressed extreme views, and it has occurred to me that a cool, detached review of the relevant codes and guides might clear the air. This then is the genesis of Sections 5.3 and 5.4.

Part I: ACI approach

Part I deals with the ACI approach, and Part II will discuss the provisions of the Uniform Building Code, no longer operational, as well as of the International Codes, both of which, despite their names, are American codes. In Section 5.4, I shall attempt to answer the question: who selects the concrete mix for residential slabs on grade for single-family low-rise houses? That section justifies the interest of Section 5.3 to concrete producers and installers.

What is a slab on grade?

Intuitively, everyone knows what is a slab on grade, also known as soil-supported slab, but vis-à-vis codes and guides an unequivocal definition is necessary.

By its very nature, a slab on grade is soil-supported, but it is relevant to consider also the loads *on* the slab. In residential single-family houses, not more than three-storey high – and this section is not concerned with any other buildings – slabs-on-grade support only themselves, and the human (as well as canine and feline) occupants and also distribute the concentrated loads of the furniture. The structural behaviour of the dwelling would not change if, instead of a concrete slab, there were a bed of gravel or an ornamental rug placed on the soil. Only comfort, cleanliness, and dryness might suffer!

The rules and practices depend much on the country in which the houses are built, and even in one country, there may be more than one design code that is in operation or that is relevant.

211

The ACI publications fall into two categories: a Code, which is mandatory when adopted by a jurisdiction, and Guides, which give advice, but which, in the form published by ACI, cannot become mandatory. Thus, there is a fundamental distinction between the two categories.

ACI Building Code

The main design code for buildings in the USA is ACI 318, whose latest edition was published in 2002. The code has a worldwide influence and is used in a number of Latin American countries. The chapter in ACI 318 headed 'Scope' says in Section 1.1.6: 'This code does not govern design and construction of soil-supported slabs, unless the slab transmits vertical loads or lateral forces from other portions of the structure to the soil.' Given that, in family houses, upper storeys are usually supported by timber frames resting on strip foundations and on isolated footings, and also given that horizontal wind and earthquake loads are resisted by timber frames, we can ignore ACI 318 as far as residential, single- or two-family houses are concerned.

The above exclusion is not new, but earlier editions of ACI 318, such as the 1995 edition, put the exclusion in the Commentary, rather than in the body of the Code; e.g. Section R7.12.1 said: 'The provisions of this section are intended for structural slabs only; they are not intended for soil supported "slabs on grade".'

ACI guides

There exist several ACI publications that relate to some extent to slabs on grade in residential construction. Being guides, they are advisory rather than mandatory. Although one of them deals with Residential Concrete (332R) and another with Design of Slabs on Grade (360R), there is no single ACI document specifically giving guidance for the design and construction of slabs on grade in single-family dwellings. Nevertheless, the provisions in the two guides are of interest, and so are some of the general statements.

I view as being of particular importance Section 2.1 in 332R, which says:

> Concrete for residential construction involves a balance between reasonable economy and the requirements for workability, finishing, durability, strength, and appearance.

Designers and professional engineers are well steeped in the need for economy and in the use of a balanced approach: they know that their task is *not* to produce the best possible structure, whatever the cost. Alas, in legal disputes, some 'experts', often backroom academics, propose superb solutions, which are unrealistically expensive. Section 2.1 in 332R should provide a powerful and convincing argument against cross-examining attorneys who say: 'It was possible to build a better structure, wasn't it?' It usually was, but only by sacrificing economy, which is not in the owner's interest.

Types of slab on grade

There is an ACI report titled 'Design on slabs in grade', 360R, which deals with six types of construction of slabs: five of these contain reinforcement, ordinary or post-tensioned, and one is a plain concrete slab. The word 'residential' is not mentioned.

A speculative digression may be in order. Section 1.1 in 360R defines the design of slabs on grade 'as the decision-making process of planning, sizing, detailing, and developing specifications generally preceding construction.' To my knowledge, such a decision-making process is rarely used in selecting slabs on grade for residential construction; rather, reliance is made on a prescriptive approach. For example, Section 2.2.1 refers to rules on joint spacing for control of shrinkage cracking.

The plain concrete slab, Type A, could be construed as being appropriate for residential use, but it is stated that: 'plain concrete slabs do not contain any wire, wire fabric, plain or deformed bars, post-tensioning, or any other type of reinforcement.' The use of the adjective 'any' – indeed used twice – precludes installing such slabs in residential construction when some reinforcement – albeit not enough to result in reinforced concrete – is fairly common.

Now, Type B in Section 2.2.2 in 360R is 'reinforced for shrinkage and temperature'. The 'nominal or small amount of distributed reinforcement [is] placed in the upper half of the slab...'. It is said – and this is important – that 'reinforcement does not prevent the cracking...'.

Section 6.1.4 in 332R says: 'Reinforcement is generally not required in concrete slabs-on-ground used for single-family residential construction. Reinforcement, however, can help limit cracking caused by drying shrinkage or large temperature changes.'

Cracking

Another important general statement is that in Section 2.3.2 in 302.1R, which says: 'Since shrinkage is an inherent characteristic of Portland cement concrete, it is normal to experience some curling and cracking on every project.' This statement should be 'writ large' so as to prevent hand wringing of experts who deplore the existence of cracks in slabs.

Nevertheless, cracking should be *controlled*. This is why Section 2.2 in 360R states: 'Since cracking is anticipated, joint spacings, usually set for crack control, are not critical, but they must be set to accommodate the construction process.' Section 2.4.1 in 360R says: 'the common objective of all of the [design] methods is to minimize cracking...'. Details of the various approaches are beyond the scope of this section, but we should note that Section 6.1 in 332R says: 'Steel reinforcing is usually not required in one and two family residential construction.'

Cracks may be due to various causes, each of which requires a different preventive approach. For instance, Section 9.1.2 in 332R states that settlement cracks can often be prevented by a proper preparation of the subgrade; and also that cracks from expanding soil can often be prevented by protecting the subgrade from absorbing water. The same section says that 'drying shrinkage cracks can be minimized by controlling the concrete mix...', and also adds: 'Drying shrinkage cracks can be held tightly by using welded wire fabric...'.

Related to drying shrinkage is curling of slabs. Advice on reducing curling caused by differential shrinkage between the top and bottom of the slab is given in Section 9.1.3 of 332R. Curling is affected by the provision of vapour barriers beneath the slab. Barriers influence moisture transmission, which is of great importance to the health of the inhabitants, floor coverings and furniture. This topic requires specialist treatment.

Quality of concrete

With respect to the actual quality of the concrete supplied to site, Section 2.2.4 in 332R contains an important statement: 'Testing of concrete is not normally done on small residential work.' The same section says: 'To verify that the delivered concrete meets the proper specifications, the purchaser may want to request a certified copy of the mix proportions.' The verb 'may' means that any decision lies with the engineer or other professional in charge.

It is also relevant – but not surprising, given that 332R is a guide – that the type of cement to be used to ensure sulfate resistance is not mandatory: Section 3.1.1 lists the cements 'recommended' for moderate sulfate exposure; for severe exposures, it says that 'Type V cement may be required'. The verb 'may' should be noted.

It is also important that Section 2.2.2 in 360R says: 'These slabs are normally constructed using ASTM C 150 Type I or Type II cement.' There is no mention of Type V cement or, indeed, of consideration of sulfate resistance.

However, Section 2.2.2 in 332R gives 'types of cement and water–cement ratios suitable for concrete resistant to sulfate attack...' in a table (Table 2.2) recognizing different levels of exposure to sulfates. That table, and very similar tables, appear in numerous guides and codes not only in the USA but also in the UK. Alas, no experimental scientific basis for such tables has ever been produced, and this bedevils a rational approach to ensuing that slabs, or any other concrete, are sulfate resisting. In any case, Table 2.2 in 332R is clearly headed 'Guidelines'.

There is one more ACI publication that deals with slabs on grade – this one at the skilled operator level: CCS-1 in the Concrete Craftsman Series. It discusses the types of Portland cement and says that Type V cement is 'used only in concrete exposed to severe sulfate action...'. The influence of that publication upon mix selection is probably minimal or negligible.

With respect to durability, Section 2.1.3 in 332R says: 'Proportioning for and achievement of compressive strength is usually assurance that such associated properties as tensile strength and low permeability will be satisfactory for the job.' Interestingly, Table 2.2 in 332R recommends for all US Regional Weathering Areas, a 28-day compressive strength of 17 MPa (2500 psi). There is, however, in Section 2.1.2 of 332R, a reference to ensuring durability in the presence of soluble sulfates, among other conditions of exposure.

Conclusions about ACI guidance

My conclusion from the above review of ACI codes and guides is that states, counties, cities, and other jurisdictions that adopt solely ACI codes, do not have any legal requirements with respect to the design of slabs on grade, unless the given jurisdiction has also made its own laws. Such a situation is satisfactory in that it gives the engineers in charge of a project the scope to make their own decision based on professional knowledge and experience.

With respect to the responsibility of the ready-mixed concrete supplier for the mix delivered, Section 4.1.2 in 332R is of importance, although, not surprisingly, it is couched in terms of 'The user should...'. This section says: it is advisable to order from a reputable and qualified

215

ready-mixed concrete producer, and to specify...'. What follows is, to my mind, ambiguous.

On the one hand, 'to specify the strength' is clear as strength can be readily verified by tests, which the producer probably performs for his own use. On the other hand, however, 'to specify...the exposure requirements...and the intended use of the concrete' shifts the decision taking on the producer. It would be presumptuous of me to criticize 332R, but in essence it says that it is for the supplier to select the mix for the given exposure requirements.

But what does the producer know about the exposure requirements? To give an example from my experience: if a jetty is exposed to the tide and splash by seawater, the durability depends on the details of the shape of the structure, of exposure to the sun and of drying conditions overall. How can the concrete supplier be familiar with these details? It seems to me that Section 4.1.2 in 332R shifts the decision-taking too far toward the supplier. But then, because 332R is a guide and is therefore couched in terms of 'should' and 'may', the guide does not impose any legal responsibility and liability.

Part II: Uniform and International Codes

Part I of this section dealt with the approach of ACI to the requirements for concrete in residential slabs on grade. Having completed that task, my reaction was: how sparse is the guidance, let alone mandatory rules on concrete in slabs on grade.

On reflection, I realized that this was probably deliberate: there is no need to have uniform, compulsory rules for non-structural elements of a house; such rules can be decided by the engineer or other professional.

However, ACI is not the only source of guidance and not the only code-writing body in the field of concrete. There exist, or more correctly existed until very recently, several other codes, the most prominent of which is the Uniform Building Code. This was published every three years, the last edition being that of 1997. The Uniform Building Code and two other American codes were subsumed in the International Building Code; there exists in fact a family of International Codes, one of which is the International Residential Code. The first edition of the International Building Code was published in 2000, and the most recent one in 2003.

Uniform Building Code

With the advent of International Codes, the Uniform Building Code became obsolete. This is so because Section 101.2 of the International

Building Code says that the provisions of the new code will apply not only to new construction but also to repair, alteration and replacement of every building. There is, however, an exception to this rule, namely 'existing buildings...shall be permitted to comply with the International Existing Building Code'.

It could, therefore, be thought that a study of the past Uniform Building Codes is primarily of historical interest; however, in litigation it is often to importance to establish whether the construction of a house complied with the Uniform Building Code *then* in force. For this reason, a brief review of the Uniform Building Code will be included in this section. The changes, over the years, in the various editions of the Uniform Building Code will not be reviewed.

In some lawsuits, the issue is whether the concrete in the slabs on grade should be made sulfate resistant by the application of specific prescriptive rules. There is no explicit answer to that question, but the 1985 edition of the Uniform Building Code, Section 2604(f) says: 'Concrete to be exposed to sulfate containing solutions shall conform to requirement of Table No. 26-A-6...'. This table is very similar to numerous tables in other publications relating the degree of exposure to the type of cement and to the maximum water–cement ratio. (A full review of this topic is presented in Section 4.3, titled 'The confused world of sulfate attack'.)

Before considering Table 26-A-6 in the Uniform Building Code as an open-and-shut case, we have to establish whether the Code applies to slabs on grade. Unlike ACI 318, the Uniform Building Code does not explicitly exclude slabs on grade. On the other hand, direct references to such slabs are very scant. Section 2623 says: 'The minimum thickness of concrete floor slabs supported directly on the ground shall be not less than $3\frac{1}{2}$ inches.' Except for post-tensioned slabs and consideration of expansive soils, no further requirements are laid down. General consideration of earthquake forces is outside the scope of this section but, it is important to note that, according to Section 2312(a), only horizontal forces above the base are relevant: the slab does not support any vertical loads and, therefore, I do not view it as being subject to earthquake design.

Exemption of slabs on grade from the requirements for structures in seismic zones 2, 3, and 4 can be found in the 1997 Uniform Building Code in Section 1922.10.3. This allows plain concrete to be used in 'non-structural slabs supported directly on ground or by approved structural systems'.

Now, slabs on grade are non-structural because Section 306(a)1 of both the 1982 and 1985 editions of the Uniform Building Code provides an exception (No. 3) to special inspections, as follows: 'Nonstructural slabs on grade, including prestressed slabs on grade when effective press in concrete is less than 150 pound per square inch.'

What I see as crucial is that parts of buildings that require no design, but are constructed following conventional requirements, are not subject to requirements included in codes in chapters headed 'The design of structures in concrete.'

Given that slabs on grade are made of plain concrete, Section 2622(a) is relevant: this prescribes the minimum ultimate compressive strength as 14 MPa (2000 psi). There is no reference to the incompatibility between this requirement and the water–cement ratios laid down in Table 26-A-6, which would result in much higher strengths. One interpretation of the situation is that Table 26-A-6 is not intended to apply to slabs on grade, but no clear-cut resolution of the situation is provided in the code. The requirement that 'plain concrete construction shall conform to the detailed minimum requirements specified in this chapter', included in Section 2622(a) in earlier codes, adds to my confusion.

The situation became clarified in the 1997 edition of the Uniform Building Code, which says in Section 1922.1.1.2: 'Design and construction of soil-supported slabs, such as sidewalks and slabs on grade shall not be regulated by this code unless they transmit vertical loads from other parts of the structure to the soil.' In Part I of this section, I showed that slabs on grade do not transmit loads from other parts of the structure.

The above clarification should be welcomed by those involved in litigation about concrete in residential housing; indeed, with clarity in the code, there should be no future litigation.

As for the practice toward the end of the last century, a survey of about 120 sets of plans for residential dwellings in Southern California, built between 1980 and 1996, conducted by Sebastian Ficcadenti, Consulting Structural Engineer, showed that all but 13 specified the 28-day compressive strength as 14 MPa (2000 psi); in those 13, the strength was predominantly 17.5 MPa (2500 psi) and occasionally 21 MPa (3000 psi).

International Building Code

This code has replaced the Uniform Building Code. Actually, there is also an International Residential Code. The latter applies to 'detached one- and two-family dwellings not more than three storeys above grade plane in height...', which are explicitly excluded from the scope of the International Building Code.

I find the situation somewhat confusing because, with regard to slabs on grade, the Residential Code contains only one section (R506.1). This lays down a minimum thickness of 89 mm (3.5 in). There is a cross-reference to Section R402.2, which specifies the compressive strength.

This paucity of requirements with respect to concrete in the Residential Code contrasts with fairly numerous references to concrete in the International Building Code which, however, seems not to apply to slabs on grade in low residential dwellings.

An interesting consequence of the above situation is the fact that the 'requirements for concrete exposed to sulfate-containing solutions' in Table 1904.3 in the International Building Code (discussed in the preceding part of this section) seemingly do not apply to residential dwellings. On this premise (which follows a strict interpretation of the International Codes), slabs on grade are not subject to any specific requirements with respect to the type of cement or maximum water–cement ratio.

I realize that my interpretation of the relation between the International Building Code and the International Residential Code may be challenged and disputes may still arise.

International Residential Code
The preface to 2003 International Residential Code describes it as a 'stand alone residential code [which] establishes minimum regulations for a one- and two-family dwellings and townhouses using prescriptive provisions.' The preface states further that the 2003 Residential Code 'is fully compatible with all the International Codes... including the International Building Code...'.

I understand the above to mean that those concerned with the construction of residential dwellings should arm themselves with the Residential Code but need not refer to the International Building Code. Nor should they expect any provisions to the International Building Code that contradict the provisions of the Residential Code. As shown earlier in this section, such contradictions *may* be thought to exist with respect to slabs on grade exposed to sulfates. I do not subscribe to such a view, but it would obviate disputes if there were greater clarity in future editions of the codes.

Given that, as far as International Codes are concerned, the International Residential Code is intended to be self-contained, the rules in it should be adequate with respect to concrete in slabs on grade.

Chapter 4, headed 'Foundations', in Section 402.2 titled 'Concrete', and in Table 402.2, prescribes the minimum 28-day strength for slabs on grade, except garage floor slabs, as 17.5 MPa (2500 psi); for garage floor slabs, the strength is 21 MPa (3000 psi) for moderate weathering potential, and 24 MPa (3500 psi) for severe weathering potential. For the condition of freezing and thawing, there are some additional requirements.

Thus, there are no requirements with respect to the resistance to sulfates in the soil. As I see it, this does not mean that no consideration of the presence of sulfates in required, but rather that any decision is in the hands of the professional responsible for the construction.

The International Residential Code uses prescriptive provision. However, engineered design in accordance with Section R301.1.3 of the International Building Code is permitted.

Other current publications

There exist other, non-mandatory publications, which could be viewed as guides. For example, the National Association of Home Builders has published a book titled *Residential Concrete*, and the Portland Cement Association a book titled *Concrete Floors on Ground*. These publications contain almost no advice on concrete for slabs on grade in residential buildings, and nothing on the problem of resistance to sulfate attack.

Conclusions about International Codes

The International Residential Code contains prescriptive provisions which (as said in Section R101.3) contain 'minimum requirements to safeguard the public safety, health and general welfare, through affordability...'. These objectives are laudable, and indeed essential in a modern society; my hobbyhorse is to make sure that affordability is not forgotten.

The provisions for slabs on grade in the International Building Code are not relevant to single- or double-family dwellings because these dwellings are governed by the International Residential Code. Alas, occasionally, the International Building Code refers to such dwellings, despite having foresworn such references in the initial part of the International Building Code. The reference may be by saying that they are excluded from a certain provision in the International Building Code;

alas, this raises doubts about other sections in which exclusion is not expressly stated.

It may be that I construe the codes in a strict manner – more strictly than other engineers would do. Alas, I am steeped in conflicting interpretations of earlier codes. Nevertheless, some clarification of the new International Codes would be helpful.

Overall, the International Residential Code is to be welcomed, and all involved in residential construction should proceed in a constructive spirit (no pun intended). Lack of too many strict rules leads to economy, which benefits the owner, but puts an onus on the construction professional to assess the situation in each case. A particular difficulty occurs when soils contain sulfates; I have written a major paper on this topic (Section 4.3 of this book).

5.4 REQUIREMENTS FOR RESIDENTIAL SLABS ON GRADE: WHO SELECTS THE MIX?

It is of great importance to concrete suppliers and installers to be clear about their responsibility with regard to ensuring that the concrete mix in slabs on grade is satisfactory, and that it conforms to the requirements of the appropriate code.

As I am not licensed in the USA, my views are those of an outsider, albeit one with a considerable experience of concrete both in the USA and elsewhere. A crucial consideration is the interpretation of the relevant codes. The situation may vary according to local custom and may depend on the existing collaboration between the design professional, the concrete installer, and the concrete supplier. Of course, in the case of design-and-build projects, all three are virtually amalgamated.

Position of the design professional

My view is that the design professional is the only person fully knowledgeable about the owner's needs and intentions as well as about the conditions of exposure of the proposed structure, and also about its intended use. All these impinge upon the properties of the concrete mix to be used. The design professional is also familiar with the relevant code or codes, and indeed he or she knows which codes are applicable and which are not.

I expect that my views are shared by other design professionals but, if they are not, other views should be aired, and the situation should be clarified. Otherwise, litigation will continue, with several parties being sued by the owner.

Interpretation of codes

A part of my argument that the design professional is the person to decide upon the concrete mix to be used is that he or she is best qualified to do so because codes are couched in terms fully familiar to the professional engineer or architect, but not necessarily to a concrete supplier or installer.

And yet, as K.B. Bondy reported in *Concrete International* in May 2004 in an article titled 'Judging building codes', a judge ruled that interpretation of the building code is in the hands of the judge. In other words, in litigation, where the plaintiff claims that the code was violated, it is the judge who, so to speak, compares the provisions of the code with the structure as built.

The rationale of this ruling is that the adoption of a building code by a city, county, or a state in the USA, makes the code into a law, and interpretation of the law is a matter for a judge, and not for the designer or for the building official. No one can argue that it should be the judge who interprets and applies the law, but the technical aspects of the code need to be viewed with full technical understanding.

The interpretation of codes is of fundamental importance in the USA because the defendants, who lost the case described by Bondy, were the suppliers of the concrete, and what was at issue was whether the composition of the concrete satisfied the requirements of the Uniform Building Code adopted by the county where the construction took place. As a layman in matters legal, but as a professional in structural engineering, I cannot see – with the utmost respect – how a judge alone, without professional advice, can interpret the code, which is written for professionals who have been trained and educated to understand the terminology such as shearing stress, principal stress, second order deformation or indeterminate structure.

In the particular case reported by Bondy, one issue was the distinction between plain concrete and reinforced concrete. Because a particular chapter in the Uniform Building Code contains 'rules and principles' applicable to reinforced as well as to plain concrete, the judge ruled that the requirements for sulfate-resisting concrete apply to all types of concrete.

And yet, the 2000 International Building Code defines reinforced concrete as 'structural concrete reinforced with no less than the minimum amounts of prestressing tendons or non-prestressed reinforcement specified in ACI 318 . . .'. The code also says that plain concrete is 'structural concrete with no reinforcement or with less reinforcement than the minimum amount specified for reinforced concrete'. The 1982 and 1985 editions of the Uniform Building Code contained a further requirement: 'and designed on the assumption that the two materials act together in resisting forces'.

The ACI 318 Code uses the same definitions as the International Building Code. Some lawyers do not appreciate this strict distinction between plain and reinforced concrete, or do not want to appreciate it. The 'sticking point' is that foundations for single-family residential homes of not more than three storeys can be built on the basis of a table in the Uniform Building Code. Although the table prescribes plain concrete, often two No. 4 or No. 5 reinforcing bars are used to counteract local lack of uniform support by the soil. By the way, foundations and slabs on grade are often placed at the same time, sometimes monolithically.

So, a lawyer can argue: the builder put two reinforcing bars into the foundations; therefore the concrete is reinforced. When being deposed, I expressed the view that dropping a bunch of keys in the concrete would thus make it reinforced: the lawyer was not amused!

Alas, amusement is a long way away from the issue of interpretation of the code. Luckily, the interpretation is not crucial in the UK because compliance with the code is not 'proof' that the design is satisfactory. For example, the British Standard 'Structural Use of Concrete – Code of Practice for Design and Construction' (drafted by a committee of which I was a member) contains the following statement: 'Compliance with a British Standard does not of itself confer immunity from legal obligations.'

The above statement means that, if a design conforms fully to a code of practice, the design is not necessarily correct and satisfactory. The converse questions: is a non-conformant design unsatisfactory? And does it render the designer culpable? are not answered by the statement in the British Standard, but non-conformant designs can be approved. Indeed, novel and 'way out' designs must be outside the scope and provisions of a code of good practice which, of necessity, codifies that which is known, has been done in the past, and is accepted.

Returning to the lawsuit reported by Bondy, the Superior Court judge's decision in favour of the Plaintiffs was appealed, but the appellants lost. However, the decision of the Superior Court has been de-published by the Supreme Court. This means that the particular decision in the given case stands, but it has not become part of the law so that it cannot be cited as a precedent in any future case. A layman would say: it was a bad decision but let it be.

What we need is a clarification of responsibilities in the selection of the concrete mix in residential slabs on grade and, for that matter, in foundations. At present, there is a lack of clarity, if not downright confusion. And litigation thrives on confusion. It is in the interest of *all* those involved in concrete to minimize litigation.

Conclusions

The issue of who selects the mix, and specifically the possible responsibility of the concrete supplier or producer for conforming to a code, is still unresolved. The responsibility of any party to the design and construction, including the building official who approves the plans, is a sensitive issue if, at a later date, litigation ensues, and a judge can himself interpret the code and thus put in jeopardy all those involved earlier in selecting the concrete mix. Codes and rules are there for the protection of the society but, unless approached gingerly, they can become weapons against some honest and competent people involved in construction.

Post-script to Section 5.4

On March 2000, in a case heard in California, Judge of the Superior Court, Hon. David C. Velasquez, delivered the following judgement:

> The Defendants [the ready-mixed concrete suppliers] were not responsible ... to the Plaintiffs to supply concrete that would satisfy the Uniform Building or any other industry standard with respect to cement type or water to cement ratio of the concrete. The design engineer and/or architect retained by [the Plaintiffs] was responsible for specifying parameters for the concrete that would satisfy applicable codes or standards ... [The] material supplier ... was responsible to furnish concrete that would meet the parameters specified by [the contractor] in its order for the concrete.

The above is of interest in that it confirms my view that compliance with design codes is the responsibility of the design professional and not of those who mix, sell, transport, place, compact and finish the concrete in the structure. After all, the codes are written by designers for designers. Acceptance of the above judgement should reduce litigation and narrow its scope.

6

General issues

This chapter looks at five different topics, not closely connected with one another but each of importance in construction.

Section 6.1 is a fairly extensive review of the processes involved in producing a structure, all the way from mix selection. The message is that, although there are numerous processes en route, each one of them is important. Good execution of every step is essential and none can be dismissed as being of minor importance. This may seem obvious while reading Section 6.1, but on site, with time pressure and with occasional labour shortages, there is sometimes a tendency to concentrate on what are perceived to be the 'essentials' at the expense of other operations. This tendency should be resisted.

The relation between workmanship and design is discussed in Section 6.2. Those who have construction site experience are well aware of the domain of responsibility of the contractor, on the one hand, and that of the designer, on the other. There should thus be no confusion between workmanship and design.

However, in the last quarter of a century, or even longer, fewer and fewer structural or civil engineers have been involved in the process of construction, let alone in the minutiae of mix selection, concrete placing and finishing, and in quality control in general. They have become increasingly replaced by scientists, many of them chemists. These people are, or become, knowledgeable about the properties of concrete and also about the procedures necessary to achieve a quality product.

That much you can learn on the job. What you cannot learn by the suck-it-and-see approach or by 'osmosis' from collaborating engineers are the basic rules of design and the response of structures to the forces and other actions applied to them. Scientists who recognize that limitation on their knowledge are most valuable on site, but trouble may ensue when they fail to distinguish workmanship from design. This is why I felt obliged to challenge the statement, 'poor design is an important aspect of bad workmanship,' which appeared in a letter in the journal *Concrete* in January 2005, page 46. I accept that, on occasion, bad design, such as a project with a poor buildability, may force the contractor to produce a poor quality structure. Good workmanship is difficult, if not impossible, to achieve, when access is poor, space is cramped, or when work has to be performed by working with the arms well above the head. In such a situation, the responsibility for poor workmanship must be laid at the designer's door. Nevertheless, even then, the contractor is not entirely helpless: he can approach the site engineer, explain the problem, and seek modifications.

But to make a blanket statement to the effect that design is an aspect of workmanship is unjustified and misleading. This is why I felt that that statement should not be allowed to pass unchallenged; that is the genesis of Section 6.2.

Another sweeping statement is the subject matter of Section 6.4. That section contains both my published letter (in unabridged form) and the original 'attack' that provoked me. A connection to Section 6.2 is that the signatories of the 'attack' are, by-and-large, not structural or civil engineers. There are no fewer than seven names of whom only two have an academic background in civil engineering. The remaining five people are scientists, and none of the seven has, to my knowledge, site experience that would enable them to opine on the application of standards and their enforcement.

'Salon generals' do not conduct their battles in the field and are therefore harmless. This is not quite the case with scientists involved in the subject matter of Sections 6.2 and 6.4. I want to be among the first who recognize the scientists' contribution to our understanding of cement chemistry and physics and to a sound approach to the study of concrete. But those who wrote the letter discussed in Section 6.4 have gone much too far. Because people without adequate knowledge of structures are willing to act as expert witnesses, it is not surprising that there is much litigation about concrete structures. I have complained about the wasteful and even pernicious nature of litigation, mainly in the USA: a similar activity is arbitration in the UK and many Commonwealth countries. My views are collected in the book *Neville on Concrete* published by the American Concrete Institute in 2003.

What is not always realized is that, regardless of who wins a lawsuit, everybody is a loser, and the final payer is the building owner – even the owner of a future building. In a nutshell, the argument is that most parties involved in construction – designers, contractors, and suppliers – carry insurance, and all the insurance premiums paid are eventually charged to the building owner. As the payouts increase, so do the premiums for future work.

Because in recent years I have been more closely involved in structural engineering, I have written Section 6.3 dealing specifically with the relevance of litigation to the structural engineer.

Finally, Section 6.5 is slightly out-of-the-way in that it considers gender in concrete. We are all increasingly being constrained by political correctness. I am not worried, at least in print, by the ethics of this situation. What concerns me, though, is the need to use clumsy or ungainly English by saying every time 'he or she' or, alternatively, the use of grammatically incorrect English by saying, for example, 'the engineer is responsible for *their* drawings' so as to avoid implying specific gender.

Gender is a linguistic concept and not a sex classification. Other languages recognize this. For example, in Latin, a sailor (*nauta*) is feminine, and the Romans certainly did not have female sailors. In German, a girl, (*das Mädchen*) is neuter, but *her* femininity is not in doubt.

Forcing a straightjacket on language can have unforeseen legal consequences. Let me give an example from my days as Principal of a University. The regulations stated: 'before proceeding to written examinations, every candidate must submit his course work'; the regulations included a blanket clause saying that 'he subsumes she'.

Now, the gender 'freedom fighters' suggested an alteration to the regulations to the effect that 'all candidates must submit their coursework before proceeding to the written examinations'. I was obliged to point out that, if a single candidate did not submit coursework (and thus *all* candidates have not done so) no candidate could proceed to the written examination. This may be a nonsensical argument but, nevertheless, the suggested new wording could lead to litigation, in which the judge might have to apply a strict interpretation of the written regulations.

6.1 CONCRETE: FROM MIX SELECTION TO THE FINISHED STRUCTURE – PROBLEMS EN ROUTE

The majority of papers dealing with concrete concentrate on one property at a time, thus giving a somewhat distorted picture of the importance of the various parameters in achieving a good concrete structure. The theme of this section is that this objective can be achieved only by a seamless process of a number of operations, which are briefly discussed. The paper concludes with a review of special aspects of precast concrete.

All studies of concrete have a single objective: achieving a satisfactory structure. Alas, despite thousands of research papers published every year, concrete in many structures is not as good as it could be. I will attempt to explain this situation.

This section is a minor provocation in the hope that by pointing out the shortcomings in concreting (and not just the concrete as such) I would shock the readers into improving the various procedures or, at least, into reviewing the present shortcomings in concreting.

So, given my good intentions, I hope to be forgiven for any harsh words that follow. And, if the cap fits...

There is one other respect in which this section differs from most others, which discuss a single topic and point out its importance. This section shows that all aspects of concrete-making matter.

Nature of the problem

Why is concrete in many structures not as good as it should be?

First, a disproportionate amount of university research is confined to purely laboratory studies on artificial specimens under unrealistic conditions, and is performed by people with no experience of what happens in real-life situations. In their experiments, the range of variables is very limited, deleterious conditions are unrealistically exaggerated so as to speed up their action, differential deformation within concrete, due to shrinkage or temperature, is avoided by the use of tiny specimens, and so on. Most academics are not renowned for investigations on an actual site or production facility, and prefer an air-conditioned laboratory to wearing gumboots and a hard hat.

Second, most civil engineering undergraduates have learned almost nothing about concrete as a material. Consequently, when involved in design calculations in subsequent employment, usually almost exclusively in computerized form, they assign fixed values to the requisite properties of concrete: modulus of elasticity, shrinkage, creep coefficient, thermal expansion, and other properties. They rarely ask themselves whether the actual mix – as yet unknown and to be supplied by a producer still to be selected – will conform to the properties assumed in the design calculations. Indeed, it is rarely that anybody looks into such conformance.

Third, overall, the labour force involved in the production of concrete, that is, batching, mixing, transporting, placing, compacting, finishing and curing the concrete is less well educated and less well trained than corresponding labour forces in many other trades, such as joinery, carpentry, electrical work, plumbing or even bricklaying. I do not mean that all concretors are incompetent: of course not; but rare is the construction site or plant where all the concretors are fully competent. And yet, any one step in the production of concrete can adversely affect the finished product. Thus, many concrete structures are imperfect with the consequence that, very soon after completion, they are in need of repair.

A small digression may be in order. The quality of workmanship depends not only on the competence of the workforce but also on the quality and intensity of site supervision. When I was young, there was a resident engineer, normally full-time, on every construction site, and his eagle eye prevented slipshod workmanship. Such supervision was expensive, but the designer's fees were large enough to cover the cost. Acute competition for design projects, coupled with abolition of fixed fees, has resulted in low fees, and the first action to save money was to economize on supervision.

Seamless process

The point of my catalogue of complaints is that achieving good concrete is a seamless project. Any one flaw will mar the end product. Sometimes the defect is visible, but at other times it is hidden and no one knows about it until something happens so that an investigation and testing follow.

In the case of reinforced and prestressed concrete structures, it is difficult to ascertain what is inside a member. In the simplest case of a highway or runway pavement, the thickness of the concrete is rarely routinely checked, although relatively recently there have been developed some electronic methods. The old system of cores is destructive. Verification of the degree of compaction (consolidation) and of the absence of honeycombing or of bugholes is not usual, if only because it is not easy, and is laborious and time-consuming. The position of reinforcing bars can be determined; provided we are certain that the correct size has been used. But even then, we do not necessarily know whether the right grade of steel was placed. I am aware of a case where colour-coding of steel was misapplied in the steel shop so that using the right colour resulted in using the wrong steel.

Low-tech concrete

Concrete is peculiar in that it is both a low-tech material and a high-tech one. This sounds like nonsense but concrete has evolved from a very simple material, which the man in the street can make without any technical knowledge, to a material expected to have very specific properties of various kinds and to be, in the engineering jargon, fit for the purpose. This development has occurred during my lifetime.

Such a development would not necessarily have adverse consequences as shown, for instance, by changes in the manufacture of cars. Less than a century ago, they were produced by skilled mechanics and assembled manually. Such cars are still obtainable today but they are worth their weight in gold! The hundreds of millions of cars available commercially every year are the product of sophisticated processes controlled by electronic means and largely executed by robots. Of course, we are not yet at the situation in a modern power plant where the staff consists of a man and a dog. Why a dog? To make sure that the man does not interfere with the system. Then why a man? To feed the dog.

Let me elaborate the changes in concrete. I remember the days when concrete was batched by hand, that is by boxes, a cubic foot in size, filled with sand or coarse aggregate or cement. Indeed, the bag of 94 lb (42.6 kg) originates from the approximate weight of 1 cu ft ($0.028\,\mathrm{m}^3$) of cement. On a small job, batching was done by shovels: so many of cement, so many of sand, so many of coarse aggregate. This is the origin of the so-called $1:2:4$ or $1:1\frac{1}{2}:3$ mixes.

The 'specification' for these materials required the sand to be clean, and rounded, and the coarse aggregate was usually gravel found in the nearest riverbed. Alternatively, if available, continuously graded aggregate could be used.

Water was added as needed, that is, so as to produce a mix workable enough to make compaction (consolidation) fairly easy. Compaction was by tamping by wooden utensils and by 'slicing' using a block of steel, about $\frac{1}{2}$ in. (13 mm) wide, attached to a rod, so as to fill the space near the surface of formwork (of course, made of timber) and especially the arises and corners.

The remarkable thing is that it all worked well: the water–cement ratio, generally neither specified nor known, was not much above 0.5, and the quality of the resulting concrete was remarkably good. Some of the elements – walls, floors, and even bridge beams – exist to this day.

I have just referred to the water–cement ratio, w/c, and this concept, developed by Abrams in the USA in the second decade of the 20th century, and even earlier by Féret in France, was known, but it was not a practical parameter on site. Indeed, in my opinion, even today w/c cannot be a primary parameter in the production of concrete. This may sound like an anathema; I will explain my reasoning in a later section.

All this was about the old low-tech concrete. Although, nowadays, we do not batch concrete by volume (with the exception of a continuous mixer), simple standard mixes containing prescribed weights of fine aggregate, coarse aggregate, and water (the latter actually batched by volume) are still used, and they produce a satisfactory concrete for many purposes. This is what I call low-tech concrete.

High-tech concrete

Now, high-tech concrete is expected to have specific properties required for a given application. These may include: a minimum compressive strength at a very early age, or a particular coefficient of thermal expansion, a low rate of heat generation (due to the hydration of cement) so that a controlled temperature is evolved, or a specific modulus of elasticity, or specific creep properties, or a limited magnitude of shrinkage under certain conditions of exposure – the last three of particular importance in prestressed concrete. This list can be extended by durability requirements for exposure to cycles of freezing and thawing, and for resistance to attack by specific external aggressive agents.

All the above can be achieved because, in the middle years of the 20th century, an enormous amount of scientific work – much of it in the USA – resulted in an understanding of the physical and chemical properties of Portland cement and of concrete made with that cement as the sole binder. In consequence, we were able to establish the properties of Portland cement that would have the requisite features. I shall consider the requirements for aggregate later on.

Precise properties of cement

Here, we come to a fundamental difficulty: we know what properties of Portland cement we require in terms of, say, C_3S, C_2S, and C_3A contents and of fineness, but can we get it? More than that: do we know what we are getting when we buy Portland cement? The answer is 'no', and this is the first conflict between expectations of high-tech concrete and reality.

Some readers may be surprised by the above assertion and point out that there exists an ASTM classification into Types I to V of Portland cement, and also a European classification into the more than a dozen blended Portland cements.

This is, of course, true but the standard classifications are very broad. For example, standard composition requirements of ASTM C150-04 with respect to SiO_2, Al_2O_3, Fe_2O_3 and MgO are the same for Type I and Type III cements. It follows that two Type III cements may differ very significantly from one another. Moreover, a particular Type I cement may have a higher content of C_3S than a given Type III cement. It should not be surprising that making two concretes with, say, a Type III cement, may result in concretes with significantly differing properties. (Type III cement is what is often used in the precast concrete industry.)

My statement could be countered by pointing out that we can ascertain the properties of the cement that we are buying. First of all, ascertaining is not the same as ordering cement with specific properties. This is possible only for very large projects, such as gravity dams and major highway projects. Indeed, according to ASTM C183-02, 'not much cement is sold on the basis of such sampling and testing'.

Second, when we ask a cement manufacturer for a certificate listing the properties of the cement that we are buying, this certificate does not apply to the cement that we are receiving on site or in the plant. Rather, the certificate tells us the average properties of the cement produced over a certain period in a given cement plant. The qualification 'average' refers to the fact that grab samples are taken in a specified manner over a period of time; they are then combined to form a composite sample representative of the cement produced during that period of time.

Moreover, it is not routinely possible to find out on which day the cement was manufactured and from which silo it originates. So, all in all, we have no more than a general idea of the properties of our particular cement.

All the above is not a criticism of the existing methods of manufacture and supply of Portland cement. Rather, it is a reflection of the fact that cement is a relatively low-cost material, and to achieve elaborate refinements in the process would result in a considerable increase in price. Moreover, the cement produced by a given plant is strongly affected by the raw materials and by the fuel used in the kiln. The fuel is relevant in so far as the sulfates in the clinker are concerned because the solubility of the sulfates influences the compatibility of a given cement with superplasticizers.

Cement-related problems

So far, I have discussed the limitation that the exact properties of the Portland cement impose on the person proportioning the concrete mix. Nowadays, Portland cement is rarely the sole binder, although it is a necessary component of the binder. There are several reasons for this.

First, by using additional binders, that is, cementitious materials (in the broad sense of the word) the properties of the combined cementitious material can be varied widely. Specifically, we are able to lower the rate of development of the heat of hydration and, therefore, of the rise in temperature of the concrete, with a resulting reduction in thermal cracking. We can also reduce the vulnerability of concrete to some forms of chemical attack. The benefits are numerous but under some circumstances, such as cold weather, too slow a gain in strength may be harmful to the desired stripping strength.

Second, many of the additional cementitious materials are either natural materials or else they are by-products in the manufacture of other materials (such as blastfurnace slag, which is formed in the production of iron) or waste products (such as fly ash, which is the fine ash formed in burning pulverized coal in a power station). Thus, these materials already exist and do not need to be manufactured expressly, with a consequent expenditure of much energy. It is the energy saving that represents an economic advantage. In addition, disposal of fly ash represents an environmental problem.

Third, it could be thought that these waste materials should be cheaper than especially manufactured cement, but often this is not the case.

234

Fly ash

Fly ash is probably the most widely used binder in addition to the Portland cement, and as such deserves special treatment in this section. I am saying 'in addition' because we must include a certain proportion of Portland cement in the binder, if only because the hydraulic action of fly ash stems from reaction with calcium hydroxide produced by the hydration of Portland cement.

The greatest protagonist of the use of fly ash in concrete is V.M. Malhotra in Canada. He has developed a methodology in which the binder contains 60% of fly ash. It is thus clear that fly ash is a major ingredient in the binder (which I prefer to call globally 'the cementitious material'); it follows that to refer to fly ash as a supplementary material is to misrepresent its quantity as well as its technical importance. Likewise, to classify fly ash as a mineral admixture is misleading: salt is an admixture to a steak and kidney pie, but both steak and kidney are the essential ingredients. Let us, therefore, refer to fly ash as a binder or, better still, as a cementitious material.

Fly ash is a particulate emission, precipitated electrostatically, in coal-fired power-generating stations. It is largely siliceous in composition, in glass form, and thus reacts with calcium hydroxide, as mentioned above. In other words, fly ash is a pozzolan. But equally important is the physical action of fly ash in a concrete mix.

The fly ash particles are largely spherical in shape, with a diameter between 1 and 100 μm; those smaller than 45 μm are beneficial. The very small particles improve packing, especially in the transition zone at the surface of the aggregate. They also reduce the flocculation of the particles of Portland cement, and thereby reduce trapping of water. In consequence, fly ash acts as a kind of water reducer.

These are the main technical benefits of using fly ash in concrete. There are also environmental benefits: if not used, the fly ash would have to be dumped, and also more Portland cement would have to be produced, with the concomitant use of energy and emission of carbon dioxide.

On the economic side, the preceding might lead us to expect fly ash to be given away free. Indeed, it was when I was a young engineer: all you had to do was to send a truck, and the power station would fill it with fly ash at no charge! Nowadays, fly ash, per ton, may be more expensive than Portland cement but, of course, it has to be good fly ash.

'Good' means that the particles are predominantly spherical, that they are appropriately small, that the carbon content is acceptably low, and that their characteristics are constant from day to day. To achieve these desirable features the power station has to use coal from the same source and the burning temperature has to be high and constant. The requirements, with respect to temperature, necessitate that the power station be a base-load station, and not a peak-load one, that is, one coming on stream from time to time. This almost means that the power station has to be connected to a large power grid.

The high emission temperature results in the emission of NO_x gases. For this reason, in the last ten years or so, health regulations in the Netherlands and some other countries have forced power stations to reduce the emission temperature, with a consequent increase in the carbon content in the fly ash and a reduction in the proportion of spherical particles. So, what the future will bring I do not know.

In the meantime, good quality fly ash is well worth paying for. The resulting concrete generally has: lower shrinkage; lower permeability, and therefore better durability; a lower rate of heat development, good pumpability and properties suitable for use in self-compaction, and good finishability. The final set is delayed by two to three hours, and the strength is developed more slowly, which may or may not matter. Finally, fly ash imparts to concrete a somewhat darker colour. The influences of fly ash on shrinkage and on the rate of development of strength are very relevant to precast/prestressed concrete.

There is one other aspect of using fly ash in concrete: it needs early and good wet curing. This instils in personnel a curing discipline. As I consider good curing to be of great importance in all types of concrete, the use of fly ash has a salutary effect: a contractor that uses fly ash on one job will acquire the habit of good curing on all jobs!

Fly ash is used extensively in some countries but not so in others – at least not yet. I view the importance of including fly ash in the mix to be so great that it merits a moderately extensive treatment in this section.

Fly ash and ground granulated blastfurnace slag have been used to produce concrete with a 28-day compressive strength of more than 110 MPa (16 000 psi).

Ternary blends

Binders consisting of three, or even more, cementitious ingredients are becoming increasingly common. In this section, I shall limit myself to the use of silica fume in addition to fly ash and Portland cement.

Interestingly, silica fume is also a waste product: it is the particulate exhaust emitted during the manufacture of silicon and ferrosilicon alloys from high-purity quartz and coal in a submerged-arc electric

236

furnace. Silica fume is always expensive because its supply is geographically limited and, therefore, transport costs are incurred.

Silica fume improves early-age performance so that its use with fly ash is advantageous. Silica fume has a high water requirement, this counteracting the improved workability induced by fly ash. Indeed, generally, inclusion of silica fume in the mix requires the use of superplasticizers. These are expensive, but a ternary mix with a superplasticizer is what makes the production of high-performance concrete possible. This is a high-tech concrete and this is the way of the future, as I see it.

High-tech concrete need not have a high strength but, to achieve consistency of a very high strength, high-tech control is necessary. This can be achieved much more easily in a precast concrete plant than on an ordinary construction site. Bridge elements, with strengths of 110 MPa (16 000 psi) or even 120 MPa (17 000 psi) have been routinely precast in France and in the USA.

Such high-strength concrete, with a very low water–cement ratio, down to less than 0.30, requires meticulous moist curing, starting at a very early age; otherwise, there is a danger of autogenous shrinkage cracking in the interior of concretes. Again, a precast concrete plant makes very good curing possible.

Mix selection

Selection of appropriate cementitious materials and their optimal combination require skill and knowledge of the behaviour of concrete in service. In fact, this applies also to other ingredients: aggregate in its various size fractions and chemical admixtures, these days more numerous and varied in their action. This selection is an art as well as a science.

All this should be done by an experienced concrete engineer. Alas, few of these are available and there has been an increasing tendency to resort to *computerized* mix designs. I am dubious about abdicating direct personal approach to mathematical formulae, if only because the properties of aggregates vary enormously from place to place, even in the same geographical area and are influenced by the crusher used.

Often, we do not know which aggregate will be used. This need not be a major problem with respect to, say, workability, but properties such as creep of concrete or its modulus of elasticity, relevant to loss of prestress, are likely to be greatly affected by the details of the mix composition.

237

It follows from the above that there would be an advantage in the designer and the precaster being linked. This, however, militates against competition in the choice of the precaster and, therefore, against economy. Many years ago, when I was involved in creep tests for the prestressed concrete containment vessels in the very early British atomic power stations; the government, which paid for these stations, applied a system of rotational selection.

For each station, the contractor was chosen ahead of the design so that, for example, creep effects could be predicted following tests using the aggregate that the contractor planned to include in the mix. To be fair, the government chose a different contractor for each project. This meant that there was no competition and, therefore, no economy. More importantly, any one contractor could not build upon its experience.

So, we have to use a performance specification, which has to be translated into a mix for a specific purpose. One difficulty is that tests are not always available. The structural engineer knows the required properties in service but he is ignorant about translating them into mix ingredients. The precaster may be able to do so but he has to build up his experience on the basis of feedback about the performance of previous mixes. Often the only feedback is a complaint.

It follows that we need sustained cooperation between the concrete producer and the structural engineer but, as I have already said, in a free-market economy, established relationships do not develop because we repeatedly look for the lowest initial cost. The trouble is that this may result in a structure that needs repairs. Of course, the answer is to use the whole-life cost but this is rare.

Sometimes, the reasons for looking at the initial cost only seem to be valid: for example, the first owner intends to sell the structure at a very early age; also, government funds are often separate for capital expenditure and for operating costs, which include repairs. It is difficult to get away from the lowest-bidder syndrome, but maybe the time has come to rethink the situation. Someone said that there are three desirable factors in construction: lowest price, best quality, and fastest completion. Seemingly, we can achieve any two of these, but not the three of them, at the same time.

Concreting operations

I have already said that w/c is not a parameter that can be insisted upon because it may be overruled by workability. I am not denying that, from the structural design viewpoint, not exceeding a certain w/c is important but if, for some reason such as a delay, the workability is inadequate, then water must be added because unworkable concrete would produce honeycombing, which is even worse than having too high a w/c.

There is a tendency to believe implicitly that the specified w/c has been implemented, but we cannot verify w/c in hardened concrete with a precision better than ±0.05. Also, according to British Standard BS 1881: Part 124, the cement content in hardened concrete can be determined at best with a difference of not more than $40 \, kg/m^3$ ($70 \, lb/yd^3$) in 95% of cases when testing identical samples of the same concrete by one analyst; this is called repeatability. With two analysts in different laboratories, the figure is $60 \, kg/m^3$ ($100 \, lb/yd^3$); this is called reproducibility. Precision is even less good if the binder contains several cementitious materials.

The only parameter that can be verified by testing is the compressive strength. Sometimes, the specification lays down the compressive strength as required by the structural designer and the w/c as required for durability purposes. At times, these may be incompatible: for example, 3600 psi (25 MPa) and 0.45. There was once a case of fraud where the ready-mixed concrete supplier ensured that the concrete had the requisite strength but used a much higher w/c than specified, knowing full well that it could not be verified once the concrete has left the mixer. Clearly, precast concrete plants routinely produce concrete with w/c of 0.35 to 0.40.

There are other reasons why w/c should not be relied upon too heavily: the allowance for moisture on the aggregate based on periodic 'snapshot' determination is not reliable, and moreover w/c varies inevitably within a structure, both horizontally and vertically, owing to segregation and bleeding.

Great improvement in the control of water content on the aggregate would be achieved by the use of in-line moisture meters; this way, the amount of water on the aggregate in *each* batch would be known and could be allowed for by varying the amount of water put into the mixer. Various types of in-line moisture meters have been developed: those measuring electrical resistance, those determining capacitance, and those measuring the absorption of the transmitted microwave signal. Alas, their use is not as widespread as it should be, but the benefits of moisture control in each batch, especially in precast concrete, are large in terms of the penalty of excessive variability of concrete; I shall refer to this in a later section.

When several mixes are used in a single structure, it is vital to ensure that the right mix gets to the right place. This is obvious, but problems may arise when different mixes are specified for different parts of foundations. This is, for example, the case with the requirement for sulfate-resistance given in the recent BRE Digest in the UK (see Section 4.3).

Such fine-tuning is good in theory but if a truckload of concrete arrives on site and it cannot be placed in the right part of the foundations, it is bound to be diverted and used in a different part, which might require a different mix. Purists can argue that the agitator truck should be sent back but this is rarely done in practice because of cost and because dumping of concrete is forbidden for ecological reasons.

In the case of precast concrete, some of the procedures, such as transportation, should not be problematic but, nevertheless, vigilance (continuous or continual) is necessary.

Finishing is a skilled operation but it should not present problems. Nevertheless, it is not uncommon for over-finishing with a steel float to produce a beautiful-looking surface, which is too rich and has too high a w/c. This may lead to cracking and to vulnerability to chemical attack, as well as to dusting.

Placing of reinforcement

It is obvious that this has to be positioned correctly but there is no unique way of prescribing the cover. Sometimes, nominal cover is specified; in other cases, minimum cover is given without a maximum value; in yet other cases, tolerances are prescribed, but it is not obvious whether they have to be observed at every point or on average over a certain distance. I have come across a case where cover in precast concrete panels with 1 in (25.4 mm) diameter bars

Fig. 6.1.1. Negative cover to reinforcement

Fig. 6.1.2. Displacement of steel during concreting

was given as 2 in (51 mm) from two sides, and the panels were only 5 in (127 mm) thick: 'negative cover' resulted (see Fig. 6.1.1). Even if the cover is correct at the instant of placing of reinforcement, the steel may become displaced during the concreting operations by the operatives standing on it (see Fig. 6.1.2).

Self-consolidating (self-compacting) concrete

Everything that I have discussed so far applies to traditional concretes that have been in use for many years. There is, however, a new development in the construction industry with a curious history: self-consolidating concrete or SCC. Like the use of robots in concreting operations, self-consolidating concrete was first developed in Japan. In both cases, the motive was to minimize the use of semi-skilled or unskilled labour, of which there was little in Japan (without allowing immigration).

Self-consolidating concrete is a mix that expels entrapped air without vibration and that travels around obstacles, such as reinforcement, to fill all space within the formwork. This is useful with intricate patterns of prestressing tendons and poorly accessible areas near anchorages. Vibration is noisy and therefore objectionable to neighbours, especially

241

at night and at weekends. Avoiding this noise is the second argument for the use of self-consolidating concrete.

There is also a third reason, and that is the health effects of immersion vibrators on operatives: holding the vibrator damages nerves and blood vessels and causes the so-called 'white finger' or 'hand vibration' syndrome. This is obviously socially undesirable, and the European Union has prepared a directive aimed at reducing the incidence of the hand vibration syndrome. It is expected that this will be translated into a law in the UK before 2006, but as yet self-consolidating concrete is not widely used in the UK; in addition to Japan, Sweden and the Netherlands are leaders in the field. In the USA, the Precast/ Prestressed Concrete Institute (PCI) already has a guide on SCC, and ACI is in the process of developing such a guide.

Because self-consolidating concrete is not well known everywhere, a brief technical description may be in order. There are three desiderata for the concrete to be classified as self-consolidating: flowing ability; passing through reinforcement; and resistance to segregation. Various tests for each of the three properties have been proposed but no *standard* tests have as yet been established although ASTM has several under development.

The means of achieving self-consolidating concrete are: use of more fines (smaller than 600 μm); appropriate viscosity achieved by a controlling agent; w/c of about 0.4; and use of a superplasticizer; less coarse aggregate than usual (50% by volume of all solids); and good aggregate shape and texture. Clearly, very good batching controls are necessary.

Self-consolidating concrete is very useful for heavily reinforced members of any shape and with bottlenecks, both in precast concrete and in situ, and also for casting concrete sculptures. The only limitation is that the top surface must be horizontal.

I am convinced that self-consolidating concrete will become more widespread in the very near future and we should prepare for it by developing appropriate mix ingredients and mix proportions.

Curing

Curing is always specified but wet curing is rarely executed well. I am a strong believer in wet curing especially at low values of w/c because water has to be added into the hardened concrete, and not simply prevented from being lost, which is what membrane curing does. With very low values of w/c, curing must start almost immediately after placing, as otherwise autogenous shrinkage will develop. The resultant cracking is not visible because it occurs within the concrete mass and this negates the benefits of a very low w/c.

I am well known for preaching about curing but this operation continues not to be done well, if at all. Many people do not believe in curing and they know that it is not possible to prove or disprove that curing had been applied. If curing were a separate item in the Bill of Quantities, it might be easier to enforce it.

Of course, fresh water must be used in curing but I have seen in the Middle East some precast concrete structures that were cured by immersion in seawater. The mistaken argument was that the precast concrete would be floated out and immersed in seawater anyway. However, at a very early age and in hot dry weather, filling the pores in concrete with seawater would facilitate penetration of chlorides into the interior of the concrete and encourage corrosion of reinforcement.

Specifying a minimum cement content

Some codes of practice and some specifications prescribe a minimum cement content, presumably because it is thought that this ensures a 'good' mix. I see no scientific reason for the belief that a higher cement content produces a better concrete.

Better concrete means strong enough and durable. As for strength, this is a function of w/c, provided full compaction has been achieved. If it has, then the cement content is irrelevant to strength. However, more cement may facilitate the compaction because more water per cubic yard of concrete can be used without exceeding the requisite value of w/c. So, it is to ensure a low enough w/c at a convenient workability that the minimum cement content is specified. Overall then, specifying minimum cement content is a means of ensuring a maximum w/c in a roundabout way.

I believe that the required w/c at the mixer can be ensured by automated controls, which will measure the water on aggregate as it enters the batcher. Methods relying upon microwaves have been developed, but their use is not as widespread as it should be. At a precast concrete plant, such controls are particularly useful as they minimize the variability of the mix.

Variability of concrete

I have dwelt on tight controls in batching and on adhering to the specified w/c. The reason for this is that the variability in the mix ingredients

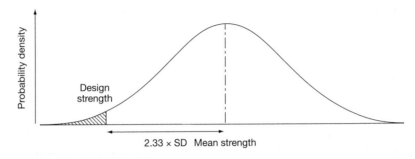

Fig. 6.1.3. Variability of strength

affects the w/c and, therefore, the compressive strength. Variability of the compressive strength means that it has a higher standard deviation, and a multiple of the standard deviations represents the difference between the average strength and the specified minimum or character-istic strength (see Fig. 6.1.3). The greater this difference the higher the cost of the mix for a given characteristic strength. This is where cost lies and this is where economies can be made.

Sustainability

Sustainability is the 'in' word in construction, so much so that this section might be deemed incomplete if sustainability were not mentioned (see Section 5.2).

The purpose of structural design is to produce structures that are safe, durable and economic. 'Safe' means that a given structure resists the loads that it is intended to be subjected to and also some unknown loads whose occurrence cannot be excluded. 'Durable' means that the given structure will function in the manner intended during the planned life (stated explicitly, for example for bridges) or during its customary life (as in the case of dwellings). 'Economic' means that the given structure is adequate for the intended purpose but not the best one that money can buy. This last limitation is sometimes forgotten but, as Neville Shute – an engineer as well as an author – wrote, 'An engineer is a man who can do for five bob (two bits) what any bloody fool can do for a quid (four dollars).'

A recent addition to the above requirements is sustainability. As I understand it in broad terms, this means that we should not use more materials or more energy than necessary so that there is something left for our great grandchildren. I have some serious doubts about this line of reasoning, particularly as in the 1960s there was the so-called Club of Rome, which predicted that, if we continued to use materials at the rate we were then doing, then they would all run out quite rapidly.

In the event, many materials such as metals in domestic fittings were replaced by other materials such as plastics. One could postulate that, when a material becomes short in supply, another material is introduced as a replacement.

The same applies to energy but, more surprisingly, every new source of energy is at least as expensive as the old one. Perhaps this is inevitable because it is the increased price of the first source that makes it economic to develop the second one. An example of the above is oil from Polar Regions whose exploration was facilitated by the increased price of oil from traditional areas.

Having said all that, it is certainly right to conserve materials and energy but, in my view, this will be hard to achieve if there is a resultant increase in cost.

Turning specifically to concrete, we should use fly ash and also recycle the aggregate. I have already discussed the technical benefits of fly ash and expressed the view that we should include it in the mix even if it is more expensive than Portland cement. To put matters into perspective, we can note that the manufacture of 1 tonne of cement produces 1 tonne of carbon dioxide.

The world-wide demand for cement is such that its production generates about 8% of all carbon dioxide produced by human activity. Another useful figure is to note that about half the carbon dioxide generated originates from transportation, and about half is produced in North America. As the Chinese and Indians *en masse* will soon be driving many millions of motorcars, a drastic reduction in the generation of greenhouse gases is unlikely unless fiscal measures are used in the present developed world.

Fiscal measures with respect to cement mean a tax on the carbon dioxide generated. As far as the recycled aggregate is concerned, some measures are already in place. For example, in the UK and some other countries there is a tax on quarried rock and there is also a tax on demolition waste turned into landfill.

It follows that converting demolished concrete into aggregate saves two taxes, as well as being a sustainability gesture, but there are some constraints. Specifically, the aggregate has to be crushed so as to have a good shape for the workability requirements and also good grading.

This may mean carting the aggregate from the demolition site to some central location. It also requires removal of reinforcing steel, which represents a resource for producing new reinforcing steel. In consequence, large-scale economic, as well as ecologically desirable, applications are limited to highways where a crusher and a classifier can move along as the old highway is 'converted' into a new one. Furthermore, especially careful control of the amount of moisture on the surface of aggregate is necessary, and the aggregate should be wet to facilitate movement of fine material. On the positive side, secondary hydration of the 'old' cement in the crushed concrete makes a small contribution to strength.

None of the above should be interpreted as my not supporting the application of sustainability criteria, if only because the production of concrete generates much carbon dioxide and uses much energy. Next to water, concrete is the most used material on a per capita basis in the world: 2 tons of concrete per capita per annum. These days, the fuel in the cement kiln can come from a variety of sources including old tires and other waste, but some of these materials produce a sulfate with a low reactivity. Therefore, for the same content of sulfur trioxide in the cement clinker, the sulfate with a low reactivity will behave differently when superplasticizers are used in the mix. This is why the compatibility of a given superplasticizer with a given Portland cement is an important issue.

Sustainability should be taken more broadly than in terms of the materials used in concrete: we should seek the best solution from the concept of a project, through design and the use of right materials, so as to produce a 'green' building. This conception includes insulation properties and also long-lasting construction. It follows that the quality of construction and the durability of structures should improve. This, then, is sustainable construction, which also has the benefit of producing less disruption through repairs. At present, in the UK, more of the construction expenditure is on repair than on new-build.

Special aspects of precast concrete

Precast concrete offers some additional benefits with regard to control of the various processes in the production of concrete.

First, the location of placing the concrete in the forms is generally fixed, so that there are no problems with a variable transportation time, and therefore, workability, or premature hydration. The quality of formwork can be superior, finishing and curing can be standardized and consistent. The quality of cement can be uniform, all of it originating from a given silo; the processing of aggregate can be standardized, both

246

with regard to grading and to moisture content. Supervision is easy to provide.

Self-consolidation makes it possible to precast elements with complex and heavily reinforced shapes. Early and consistent wet curing can be applied, and so can steam curing.

These days, very large and heavy units can be precast and transported; in the case of bridges with a double curvature, 'mating of surfaces' can be provided by using a preceding element as a 'formwork' surface.

Very high strength concrete can be achieved and the concomitant need for very early curing can be satisfied.

The list of advantages of precast concrete is long. In summary, most of the needs for a seamless process are brought into one location, and this is conducive to high and uniform quality of precast concrete.

While the contents of this section are applicable to all concrete – whether precast or in situ – there are three aspects of particular relevance to precast concrete. The first of these is the considerable advantage of including fly ash in the mix. Second is the meticulous control of moisture content on the aggregate that can be ensured by in-line moisture meters. Third, and last, proportioning the mix so that the concrete is self-consolidating bestows particular benefits in the process of precasting.

Concluding remarks

This section is not a denigration of concrete: how could I belittle concrete when it has fed and watered me over more than 50 years. (The verb 'water' should not be taken in too restricted a sense.)

The preceding pages have been a high-speed review of the various activities and decisions necessary to produce a satisfactory concrete structure. Of necessity, no more than *the* barest discussion has been possible, and of course an intensive discourse was not planned. Nevertheless, I hope that I have shown the multiplicity of activities and decisions involved, and the importance of each of them being correct and well executed. This is the only way to achieve a satisfactory structure: ignoring or treating lightly any one step ends up in a deficient end product.

Alas, the latter is not uncommon: we must do better!

6.2   WORKMANSHIP AND DESIGN

All those involved in construction are aware of the fact that workmanship and design are quite distinct, especially in terms of responsibility. Structural design is the exclusive province of the designer, that is, the engineer; the only exception is temporary works, which are usually designed by the contractor but the design has to be approved by the engineer. On the other hand, workmanship is the quality of Works that is assured by the contractor, but the engineer has to ensure that the workmanship is of appropriate quality.

There should therefore be no confusion between design and workmanship, but workmanship is not entirely independent of design. These days, much attention is given to so-called buildability; this means that the structure can be constructed without unnecessary difficulties. Moreover, the designer must pay appropriate attention to health and safety aspects of construction, and must not expose the labour force to avoidable hazards.

The engineer's responsibility for workmanship is embedded, for instance, in the ICE Conditions of Contract [1], which say in clause 36(1): 'All materials and workmanship shall be of the respective kinds described in the Contract and in accordance with the Engineer's instructions. . . .'

In commenting on clause 36(1), Eggleston [2] says: 'On workmanship, the general rule is that the contractor is to use reasonable skill and care. Thus section 13 of the Supply of Goods and Services Act 1982 says: "In a contract for the supply of a service where the supplier is acting in the course of a business, there is an implied term that the supplier will carry out the service with reasonable skill and care".'

All this is well understood by engineers but, these days, scientists are much involved in construction and sometimes they may be less clear about the distinction between workmanship and design. Hence, to say without qualification that: 'poor design is an important aspect of bad workmanship' [3] may send wrong signals to those involved in construction; it is therefore useful to review the situation.

An observant reader of my paper will have noted that I have not defined or described workmanship; nor have the ICE Conditions of Contract [4] done so. The situation is akin to the comment that it is easier to recognize an elephant than to define it. In other words, we know poor workmanship when we see it.

Design codes also use the term workmanship without defining it or even explaining it. For example, CP 110:1972 [4] contains several sections with the word 'workmanship' in the title but only specific requirements with respect to concrete or reinforcement are given. Likewise, in BS 8110:1985 [5] there are two sections headed workmanship, but the term does not appear in the text. BS 8000: 1989–2001 is entitled Workmanship on Building Sites [6], but the term is not defined in the text.

The *Designer's Handbook* to CP 110:1972 [7] limits itself to saying: 'The workmanship shall be of the quality specified, and all persons employed on the Works shall be competent and skilled in their respective occupations.'

In his book on Construction Law, John Uff [8] refers to clause 36(1) of the ICE Conditions of Contract [2] by saying that it 'should be read in the light of the general obligations to comply with the contract...'. Thus complying with the requirements of the contract, in so far as workmanship is concerned, is the contractor's duty. On the other hand, the contractor has no responsibility for design.

The distinction between workmanship and design is clear when the designer is a separate entity from the contractor and the structure is fully specified, as is the case in buildings or many bridges. It was my article on buildings made of precast units that Bensted [3] referred to. On the other hand, as pointed out to me by Sir Alan Muir Wood, when there is a significant uncertainty about physical conditions, as in tunnelling, the designer needs to ensure that assumptions are related to what is found in place so that exceptions can be referred back for action. Also, definitions of the areas of responsibility are eroded when the contractor is in charge, with the designer being an employee.

Strangely enough, court rulings are not helpful in defining workmanship. Expressions such as 'in a workmanlike manner' abound. For example, in the oft-quoted case *Greaves & Co. v. Baynham Meikle*, Lord Denning, Master of the Rolls said in his judgment in Appeal: 'It is a term implied by law that the builder will do his work in a good and workmanlike manner...' [9].

Interestingly, the absence of a legal definition of a term used in law is not unique. These days, much is reported about fraud, and in the UK we have a Serious Fraud Office, and yet English law does not provide a definition of fraud, nor is there a substantive offence of fraud at criminal law.

It may be useful to quote from the *Shorter Oxford Dictionary*, 1993, the definition of workmanlike: 'in a manner or style characteristic of a good workman; competently'. Also, to use modern parlance, court rulings refer workmanship to a 'benchmark' of the kind: quality of work of a competent contractor specialized in the given field.

There is a subsidiary question: are workmanship and the quality of materials distinct? Lord Denning distinguished 'work done in a good workmanlike manner' and 'a supply of good and proper materials'. Likewise, Lord Diplock distinguished 'the way in which the work was carried out' and 'the materials'. On the other hand, the *Building Contract Dictionary* defines workmanship as 'skill in carrying out a task', and differentiates between things (goods and materials) and 'the work done on them to produce the finished buildings'. My own view – that of an engineer and not a lawyer – is that much depends on whether the quality of materials was expressly specified in the contract (e.g. timber of a certain grade). If not, workmanship covers all aspects of fashioning an artefact.

Anyway, we all know what is poor workmanship, and the user of the building knows when to complain. But design is another kettle of fish: the strength of a structure in relation to design loads is not apparent to a naked eye; nor is robustness. Serviceability and durability can be assessed by a professional engineer but not just by eye.

The contribution dealing with workmanship also states that the chemical and engineering (or structural) properties of high-alumina cement are 'two heads of the same coin' [3]. In my view, they are not: at most, they are complementary. In any case, the chemical properties refer *primarily* to cement while the structural properties refer almost exclusively to concrete made with the given cement. I recognize the important role of cement chemists but not as structural designers. There is a parallel here with the saying: a structure designed by an engineer without an architect is horrifying, but a structure designed by an architect without an engineer is terrifying.

References

1. *The ICE Conditions of Contract: Seventh Edition*, ICE, London, 1999.
2. Eggleston, B., *The ICE Conditions of Contract: Sixth Edition – A User's Guide*, Blackwell, Oxford, 1993.
3. Bensted, J., High-alumina cement in UK construction, *Concrete*, 39, 1, Jan., 2005, p. 46.
4. British Standard CP 110:1972, *The Structural Use of Concrete*, BSI, London, 1972.
5. British Standard BS 8110:1985, *Structural Use of Concrete*, BSI, London, 1985.

6. British Standard BS 8000:1989–2001, *Workmanship on Building Sites – 16 Parts*, BSI, London, 2001.
7. Reynolds, C.E. and Steedman, J.C., *Reinforced Concrete Designer's Handbook*, Viewpoint Publications, London, 1974.
8. Uff, J., *Construction Law: Law and Practice Relating to the Construction Industry*, Sweet & Maxwell, London, 1999.
9. *Greaves & Co.* v. *Baynham Meikle*, Court of Appeal, 14 May, 1975, *Lloyd's Law Reports*, 1975, Vol. 2, p. 326.

6.3 RELEVANCE OF LITIGATION TO THE STRUCTURAL ENGINEER

An argument for the relevance of litigation to the structural engineer is the recent publication by the Institution of Structural Engineers of the second, substantially reviewed, book on expert evidence [1], and also the publication of an extensive review of that book by Pepper [2]. It is these publications that have provoked me into sending the present section to *The Structural Engineer*. It is relevant that about one-half of my experience as an expert witness has been in the USA. While I am not making any explicit comparisons between the British and American approaches, the latter may well be of interest in these days of world-wide engineering activity.

Of course, many structural engineers have never been in a courtroom, and their activity is design in an office, construction on site, or research in a laboratory. Yet, litigation touches us all because it affects the good name of construction. To look at it another way, what all of us have in common is an interest in better design and construction, in my case using concrete, and in the use of better concrete in construction. Not satisfying any of these desiderata may lead to litigation and, whatever its outcome, it has an adverse effect on the image of concrete, and hence does not promote its use.

As I see it, therefore, we should minimize litigation, and this is best achieved by reducing, if not eliminating, those factors that may give rise to litigation. In the broadest terms, if we build structures about which no party need, or indeed can, complain, we shall have, in the words of Voltaire, 'the best of all possible worlds'.

I believe that publications about concrete, including the present book, are an appropriate medium for promoting concrete construction that will not be subject to litigation. Yet, many construction projects, especially those involving concrete, end up in litigation or arbitration; as both these arise from similar contracts and as both are adversarial in nature, I shall use the term litigation to subsume arbitration.

At this juncture, I cannot resist telling an old anecdote about there being four legal 'systems' in the world. In the British and American systems, everything that is not expressly forbidden is allowed. In the Swiss system, everything that is not expressly allowed is forbidden. In the French system (my favourite), even that which is forbidden is allowed. In the old Soviet system, even that which was expressly allowed was forbidden.

International arbitration

In May 2004, I organized a seminar on 'Litigation issues in the concrete industry' at the CANMET/ACI International Conference in Las Vegas; the organizing body is Canadian and American, but the audience is world-wide. There were five presentations: one by a partner in an American firm of attorneys; one by a registered arbitrator in England; one by an executive vice-president in a large engineering company in Canada whose activities include construction materials, building science and testing; one by a petrographer heavily involved in problem solving; and the fifth one by a principal in a large engineering office in the USA. These five people wrote their contributions independently of one another. Their papers are of general interest and are of generic interest: they are not about any lawsuit, current or past, although of course the authors' opinions are inevitably influenced by their experiences. Anything else would be theorizing.

There are cultural differences between different countries in construction and in litigation: while in the USA it is almost a rule that all parties to a contract are American, this is not so in many other parts of the world. For example, I recall being involved in a very major development project in the Arabian peninsula where the client was Middle Eastern, the designer British, the firm that won the construction contract was Spanish, and the largest subcontractor Korean, whose many technical employees were from the Philippines, and operatives from India and Bangladesh. The only thing they had in common was, that *lingua franca*, 'bad English'.

Why is there so much litigation?

First of all, widespread litigation is not limited to construction. In the UK, for example, any mishap in surgical procedures may lead to litigation with a claim for large sums of money not only for a surgeon's clear mistake, but also for a somewhat unsightly scar, for consequent shyness of the 'sufferer', and for the accompanying mental distress. Lawyers who follow accidents have given rise to Charlie Chaplin's portrayal of an 'ambulance chaser'. If a holiday brochure promises a better view from a bedroom than there is in reality, this can lead to a claim not only for the moneys paid out by the holidaymaker but also for his 'suffering' through loss of amenity.

Justice, in the strict sense, is sometimes not the object of litigation. There is a story about a party to a dispute who had to be away when the judge delivered his decision. His lawyer, on winning the case, cabled his absent client: 'Justice prevailed'. The client's reply was: 'Appeal immediately'.

In England, very many highway contracts result in arbitration, and provision for arbitration is embedded in most contract documents. The same applies to a great many standard contract documents for a wide range of construction projects.

In the USA, where litigation and not arbitration is the preferred, and almost universal, route to seek redress, there are many court cases involving housing construction projects, in which developers, contractors, and subcontractors are all pursued for alleged faults and failures, sometimes by batches of 100 or more homeowners at a time.

All these examples are a manifestation of the litigious culture in our societies. Regardless of the outcome of any given court case, the cost is likely to be very large in the time spent by all the parties, the fees of lawyers on all sides, and the honoraria of expert witnesses, not to mention the charges for tests on materials.

Side consequences of litigation

One reaction to this could be: so what? Let the culprit pay! But it is not really the 'culprit' who pays; it is his insurer. Of course, the money does not come out of the insurer's pocket, but from the insurance premiums paid by the designer, the contractor, and everybody else involved in construction. And these insurance premiums are part and parcel of each party's costs, which, of course, ultimately come out of the client's, that is the owner's, pocket. So everything is dearer than it need be.

This, however, is not the end of the story. As the sums spent on litigation rise, so do the premiums and eventually some activities cease to be viable or even possible. Let me give some examples. In Australia, some insurers refuse to insure obstetricians either altogether or impose a claim limit that is lower than a possible financial award to the plaintiff. In consequence, some doctors retire early, others move to a state where there is a statutory ceiling on damages that can be awarded, and that is acceptable to insurers and commands an affordable premium.

Where a reasonable solution to the cost of insurance cannot be found, for example by a statutory limit on claims, the consequences are to the detriment of those who were supposed to be protected by compensation. For example, doctors may charge their patients very high fees, which are not affordable to people with modest incomes. So, instead of being helped by law, these people are deprived of medical care and either get none or have to travel over a state line to a 'cheaper' jurisdiction.

What is the relevance of all this to construction? First, if construction is *perceived* as a more risky business in financial terms, then entrepreneurs will be less likely to be involved in it, and will invest elsewhere. More specifically, if use of concrete is *perceived* as a more risky activity, then it may be used less than, say, steel. To my knowledge, there is less litigation about steel construction than about concrete, possibly because it is easier to 'attack' concrete for the mixture as selected or as used, or for workmanship in terms of local imperfections. Steel is factory made and its properties are indisputable. And yet, those of us 'in concrete' know that concrete structures are as good as steel structures and even offer some specific advantages, for example, in terms of fire.

Causes of litigation in construction

The litigiousness of modern society in some countries is not an adequate explanation of the situation. After all, you cannot seek legal redress without a cause, or at least an alleged cause. In the case of construction, there are two probable causes: bad construction and bad contractual arrangements.

As for bad construction, there is no excuse for an unsatisfactory design or for improper execution. Admittedly, there are occasions where some conditions or events were not foreseen or could not have been expected. In some other cases, provision for extreme conditions was not made because the probability of an event is so low that prevention would be unreasonably costly. This applies, for example, in not considering a simultaneous occurrence of maximum live loading, such as snow, maximum wind loading, and earthquake loading. In some areas, we make no provision for hurricanes or freak floods; thus the construction is not really 'bad', but involves a voluntary risk. Unfortunately, the owner is not always fully aware of the situation.

Now, to turn to bad contractual arrangements. I am using the term 'contractual arrangements' to include not only the main contract document, but also subcontract specifications and the drawings. The main contract is usually a well-prepared document, but it sometimes puts too heavy an onus on the contractor, often couched in terms such as: 'the contractor shall satisfy itself'. Also, the time allowed for construction may be unrealistic. These over-onerous requirements may give rise to a claim.

255

More often, problems arise from a specification that is unclear or ambiguous or internally inconsistent. A common reason for this is the preparation of a specification by a 'scissor-and-paste' method or computer stitching, that is, putting together elements from existing specifications for other projects, which are different in nature, without reviewing the elements for appropriateness, applicability, practicality under the new circumstances, or consistency with other clauses. I remember seeing in Hong Kong a clause about curing to prevent frost damage!

An experienced contractor, especially a wily one, will spot weaknesses in the specification that can be exploited. I remember a case in an English-speaking country where the contractor was German. During the trial, it was obliged to disclose its documents used during bidding, and there, lo and behold, was a comment in the margin in German: 'good point for a claim'. So it was, because the specified requirement was not sensible; I spotted the comment because I have some knowledge of German.

This case begs an interesting question: was the contractor guilty of dishonesty? No. Was he underhand? Maybe, but this is no crime. Or maybe he felt that, to win the contract he had to put in a successfully low bid, but to make a profit he had to find a way out. The policy of choosing the lowest bidder, regardless of anything else, may, in the long run, turn out to be counterproductive.

Personal relations

There is another factor in the propensity to litigation that I have observed in some cases in the UK. This is the relation between the engineer representing the designer on behalf of the building owner on the one hand, and on the other, the contractor. Under the contract, the engineer has to approve quite specifically much of what is done on site, and he also has the power to direct the contractor to proceed in a particular manner, or even to remove work deemed to be unsatisfactory. This has to be well documented.

All this is as it should be, but sometimes the correspondence between the engineer and the contractor becomes progressively more acrimonious and the verbiage increasingly abrasive. At that stage, it is difficult for parties to have a cool discussion aiming at a mutually acceptable solution; threats are made and, eventually, the parties proceed to arbitration or litigation. Once this path has been taken, no concessions can be made, lest they are interpreted as a sign of weakness.

I have to add that, in the UK, steps have recently been taken to obviate the above problem. Many contracts now require the appointment of an adjudicator at the beginning of construction. He or she is, so to speak, waiting in the wings and, as soon as there is a disagreement, and before it has developed into a formal dispute, the adjudicator steps in and takes a provisional decision. This allows the work to progress unhindered by soured relations. Later on, if either party is dissatisfied with the adjudicator's decision, it can refer that decision to a tribunal.

I have described the disputes that are peculiar to the UK and to many Commonwealth countries. In some other countries, fewer problems arise because there is a less sharp division between design and construction: many projects are of the design-and-build type, so that the designers are embedded in the contractor's organization. There are also cultural differences. Although my experience in France is very limited, I believe that in that country it is often the case that the designer and the contractor each employ a professional engineer with the same educational background. The two engineers approach the problem in the spirit of professional colleagues, rather than in terms of the dollar, or perhaps I should say euro, sign.

Arbitration

Arbitration, although it deals with a fully blown dispute, is also meant to resolve the problem from the professional as well as the contractual standpoint. The practice of arbitration, which goes back to the middle ages, involved a specialist who gave his decision on issues such as: has the grain spoilt while being shipped? This specialist was chosen jointly by the parties to the dispute by reason of his specific expertise in the given field.

In the last half-century, with more complex legal rules, the arbitrator could, increasingly rarely, take a decision *ex aequo et bono*, that is, on the basis of what is fair and just, but was obliged to conform most carefully to legal provisions. This change opened arbitration to lawyers; when I was a Fellow of the Chartered Institute of Arbitrators, I found that more arbitrators were lawyers or quantity surveyors than engineers and architects. Even retired judges were there to offer their services in dispute resolution. It is therefore not surprising that the proceedings centre on legal issues.

Technical presentation in court

I hope the preceding remarks will not lead to an accusation of my being anti-lawyer; if nothing else, the fact that my father was a legal counsel would not allow me to have such an attitude. Nevertheless, I am somewhat unhappy about the need for technical matters to be presented by lawyers.

I suppose the origin of lawyers addressing the judge on behalf of the client arises from the fact that, in the past, the lack of literacy and education of some people in the dispute meant that they were unable to present their opinions to the judge and jury in an adequate manner: the lawyers were there to help the lay people. Nowadays, this lack of ability of lay people to present and explain matters rarely occurs with experts. What concerns me is that, in the common-law system, an expert witness is usually unable orally to address the tribunal direct, but can only answer questions put to him by the lawyer for either party to the lawsuit. This may not elucidate all the relevant opinions if the witness's lawyer does not ask a question that would enable the witness to deliver the crucial information.

In a cross-examination, the lawyer, who after all is not, and is not expected to be, a technical expert, may fail to ask a clinching question of an expert witness for the opposing side. In England, there is a clumsy remedy for this: on a number of occasions, I have been asked by the lawyer who engaged me to sit behind him and to feed him notes with questions to be asked of the witness being cross-examined: my technical knowledge was intended to help to compel the witness to expose weaknesses in his evidence. There is no impropriety in this; indeed, the lawyer sometimes momentarily stopped the cross-examination and conferred with me in whispers to find out what was a 'good' question to ask. What a charade! My problem was that I am rather deaf. The more general problem is that the proceedings are occult in the original sense of the word, that is, concealed if not mysterious.

I am probably naïve, but perhaps a public discussion between experts would more quickly and more fully bring forth the technical facts. On reflection, I *am* naïve because such an approach would put a premium on expert's skills in rhetoric and advocacy. So, I offer no quick remedy.

What can we do?
Some readers will dismiss the above as 'here is an old man being critical of the system that feeds him'. I am old, but to be convincingly critical you should have experience.

I do not think I am alone in complaining that we live in an increasingly litigious society, with ever-growing monetary compensation. This allows human greed to grow and also allows the greed to be fostered by sometimes unscrupulous lawyers. Contingency fees, common in the USA, contribute to this, although I recognize that they enable the 'little man' who is hard done by, to seek redress in law.

For expert witnesses to work on a contingency basis is inexcusable. They must not do it overtly, but they may be told that re-engagement in the next case is contingent on what they say in this one.

Criticism of the existing system is becoming louder. A cartoon recently showed a lawyer talking to a client about an asbestos claim, a popular claim about existing buildings since the noxious health consequences of asbestos were established. The lawyer says: 'This is usually an open-and-shut case. I open my safe, put the money in, and shut it.'

In the construction world, we are not being helped by the growing complexity of legal obligations: guarantees, warranties, exclusions, penalty clauses, joint responsibility and so on. Lawyers are good at preparing the relevant documents, and lawyers are also the largest profession in the legislature in the UK and in some other countries. They are there because they are able, but perhaps we can hope for *able* engineers to make their voices heard too. If we, engineers, succeed even a little in reducing litigation, we shall progress because 'nothing succeeds like success'. I have no illusions about a rapid change but I hope that airing the present problems might be of some help: to be forewarned is to be forearmed.

I have written on the issue of impartiality and reliability of expert witnesses in the past, both in the USA and in this country, but this is not my *idée fixe*. As recently as 5 September 2005, there came into force in England and Wales 'The Civil Justice Council Protocol for the Instruction of Experts to give Evidence in Civil Courts', approved by the Master of the Rolls. This is to help expert witnesses and lawyers instructing them to comply with Civil Procedure Rules published, as recently as 1999, by Lord Wolf, the then Lord Chief Justice.

The Civil Justice Council Protocol reiterates that experts should 'provide opinions only in relation to matters which lie within their expertise'. This is hardly new.

The crucial problem addressed by the Civil Justice Council Protocol is the conduct of expert witnesses in terms of veracity, independence and impartiality. It is easy to theorize that the expert witness should be able to confirm that he or she 'would express the same opinion if given the same instructions by the opposing party'. It is a pity that it is necessary to reiterate 'the overriding objective that the court deals with cases justly' and to state that 'they have an overriding duty to help the court on matters within their expertise. This duty overrides an obligation to the person instructing or paying them.' In crude language, the experts' opinions should not be for sale.

In a city renowned for corruption, there was a boast: we have the best mayor that money can buy.

Having heard all the above, Sherlock Holmes would have exclaimed to Watson 'Elementary', but alas it is not. In the 1 November 2005 issue of *The Structural Engineer* there is a legal column, entitled 'Can credibility of expert witnesses be restored?' It is sad that credibility needs to be restored and even sadder that the question has to be put.

The issues of litigation and justice, honesty and truthfulness are not new. In *Iolanthe*, nowadays called a musical, first produced on 25 November 1882 and still popular in England, W.S. Gilbert wrote:

> 'When I went to the Bar as a very young man,
>   (Said I to myself said I)
> …I'll never assume that a rogue or a thief
> Is a gentleman worthy implicit belief,
> Because his attorney has sent me a brief,

And later

> I'll never throw dust in a juryman's eyes,
> Or hoodwink a judge who is not over-wise,
> Or assume that the witnesses summoned in force
> …Have perjured themselves as a matter of course,
>   (Said I to myself said I.)

We are still awaiting a musical about construction litigation.

References

1. IStructE, *Expert Evidence: a Guide for Expert Witnesses and their Clients*, Second Edition, 2003.
2. Pepper, M., Expert evidence guide, *The Structural Engineer*, 4 May, 2004, p. 10 & p. 13.

DISCUSSION

Letters in the 'Structural Engineer'

On 4 January 2005, David Quinion wrote

Adam's paper effectively describes the intrusion of litigation and its negative effects on our activities as engineers in the construction industry. There are three other factors we should not overlook in reviewing litigation as they are rarely discussed in this context. The first concerns the status of engineers. After World War II there were few engineers employed by contractors compared with consulting engineers who were undertaking the designs for clients. The major consultant practices (mainly with names that rolled off the tongue) administered the construction contracts on behalf of their clients and dealt with variation orders and claims from contractors in an impartial manner interpreting the requirements of the contract and specification fairly for client and contractor alike. For the contractors the system worked well as they dealt with the engineer who they met regularly and with whom they usually established reasonable working relationships. In *impasse* situations there was a senior partner or arbitration under the contract. Resolution was usually quick and often recognized that the parties needed to work together in future.

Within a decade this independent role of the consulting engineer began generally to crumble as clients intervened. They required the involvement of administrative, financial and legal staff to deal with or amend the standard conditions of contract and other requirements, including their own interpretation of them. Too often they chose, and unwisely appointed, inappropriately qualified consultants and contractors. At the same time, the industry was being subjected to a succession of new and amended statutes, measures and controls which increased the costs of operation. The viper of litigation had much to feed upon! The second factor was the realization by responsible organizations of the need for better and more relevant conditions of contract, specifications and supporting technical and performance guidance [therefore hopefully resulting in] less uncertainty, delays and litigation. These would not only encourage the adoption of good working standards and practices but enable the elimination of organizations incapable of complying with them. This required money for more staff training and the encouragement of experienced staff to participate in nationally recognized committees formulating these recommended documents. The shared views of committee members agreed new documents and interpretations about matters which on individual contracts had been disputed by the opposing interpretations of two of those members for somewhat different reasons. Such committee work would lead to support for research bodies such as CIRIA. These improvements were hampered by clients who skimped on the early costs of soil surveys, other pre-contract information findings and appointed inappropriate people. The recent and present emphasis on price cutting seems to result in reduced spending on staff development and reduced use of experienced staff for collaborative committees. The third factor is that a means of reducing litigation lies in the correct use of audited quality management systems (QMS). The use of such systems gives assurance that the finished work will demonstrably conform to previously agreed contractual conditions and specifications. It is important that regular audits demonstrate the full effectiveness of the system in use and that the staff employed are appropriately trained and experienced, as possibly demonstrated by their records of CPD, for the duties allocated to them. The consulting engineers thus define and agree their briefs with their clients and agree construction proposals with the appointed contractors. Each party knows the what, why, when and how of their separate commitments. Properly trained and experienced staff of both the design and the construction teams should understand the demands of the project from both design and construction viewpoints and so collaborate for successful projects. There should consequently be no grounds for dispute and litigation to waste time and money and taint future relationships.

On 1 February 2005, David Doran wrote

Adam Neville's interesting article 'Relevance of litigation to the structural engineer' (*The Structural Engineer*, 5 Oct. 2004) re-ignites the whole argument about the climate within which engineers and other professionals have to pursue their work.

The construction industry has available to it a cascade of solutions to a dispute resolution. In order of severity [and cost] these are:

- negotiation
- mediation
- adjudication
- arbitration
- litigation.

For some years I supported the late Kenneth Severn, Peter Campbell and others in the formation of the Construction Disputes Resolution

Group (CDRG). This offered a speedy, professional, inexpensive and effective method of dispute resolution. CDRG had some significant successes but the take-up was limited.

Our Parliament is, in my opinion, overladen with lawyers and those with little experience of business. Is it too naïve to suggest that legislation be enacted that all construction disputes should be settled by following the above cascade before reaching the courts? In that way, most disputes would be satisfactorily (and inexpensively) settled long before reaching the courts.

The introduction of an adjudication clause into some contracts is a step in the right direction but, in my opinion, does not go far enough.

The above led to a following comment from the Journal
The difficulty is that parties to action can become entrenched in their views with insufficient will to accept less than their dues suggested by their lawyers or advisors. The CDRG should be more popular than it is and if accepted by the profession should produce, as Dr Doran says, a significant saving in costs.

A letter from Sir Duncan Michael to Adam Neville, dated 14 October 2004 said
I liked your article about litigation and engineers. My direct experience is limited but my few experiences have left strong memories. I deeply dislike litigation whatever its technical form. That is an emotion, though I can explain it more or less rationally. There are various subspecies of litigation, like mediation, but they all arise in the same context of a failure of a relationship. It is not the alleged error that is the problem; it is the failure to agree how to put it right.

POSTSCRIPT ON MEDIATION

Earlier in this section there have been several references to mediation as a mechanism for dispute resolutions. My personal experience in Australia, about which a note was published in the *Journal of the Chartered Institute of Arbitrators*, may be worth mentioning.

There is no unique definition of the role and power of the mediator, and it seems to me that the rules need not be fully spelled out in a general manner but can be left to an *ad hoc* agreement by the parties and the mediator. Since the acceptance of any solution, that is the resolution of the dispute, is entirely at the discretion of both (or more) parties and, once reached, cannot be subsequently revoked or challenged, what matters from the outset is the likelihood of success. The prerequisite for this is the willingness of both parties to a find a mutually acceptable, or equally disliked, solution. It follows that what is crucial are the attributes and the performance of the mediator.

But no single man or woman may possess all the desirable attributes, and I was impressed by the very successful performance of the two co-mediators, one of whom was an eminent lawyer and the other an experienced engineer. It did not matter the that two did not always share a viewpoint or an opinion – a situation which would render arbitration impossible – because neither of them, or even both, could impose any decision on the parties to the dispute. The tremendous advantage of the mediation 'team' was that it was equally able to interpret the legal aspects of the contract and to understand the technical issues.

The procedure was simplified and the proceedings shortened. There was, by agreement, no legal representation of either party, and this was beneficial in avoiding side issues. Equal and fixed time for presentation was given to both parties, and witnesses could be stopped short. One witness was treated to a priceless remark by the mediator: 'your over-enthusiasm may diminish your posture of objectivity.' The parties set themselves a deadline to resolve their multi-million dispute: three days of hearing, interspersed by negotiation (following of course extensive written exchanges), failing which they would proceed to arbitration. The prospect of that drawn-out process as well as of a considerable delay in any money changing hands (important at a time when cash flow loomed large and interest rates were high) had the effect of concentrating the parties' minds almost as wonderfully as, according to Dr Johnson, is the case when a man knows he is to be hanged in a fortnight.

The limitation on the totality of time spent on each party's witnesses was highly beneficial; the procedure is very simple, relying on a chess clock. The time limitation can, of course, be applied also in arbitration because it is the parties who control the time, and not the arbitrator who could be accused of bias.

6.4 VIOLATION OF CODES

This section follows logically Section 6.2 and deals with a kindred attack on structural engineers. The 'attack' was in the form of an unsolicited letter in *Concrete International*, published in October 2003, pages 17 to 19, which is headed 'Chemical attack on concrete'.

There are seven authors of the letter led by Jan Skalny: not one of them is a licensed and practising structural designer, and five of them are chemists.

What I consider an almost gratuitous attack is preceded by statements to the effect that 'we know how to make durable concrete' and that 'It cannot be overemphasized that the problem now is not lack of knowledge.' Unfortunately, this is followed by: 'To the contrary, the problems are inadequate adherence to the best practices of concrete making, routine violation of the existing standards and lack of their enforcement, and our continuing inability to transfer the existing knowledge to the engineering practice.'

Adam Neville's letter in *Concrete International*, March 2004

In the October 2003 issue of *Concrete International*, there is a letter making an assertion that should not be allowed to remain unchallenged because it impugns the good name of concrete, which ACI promotes so hard. Skalny and six others wrote: 'the problems are inadequate adherence to the best practices of concrete making, routine violation of the existing standards and lack of their enforcement...'.

This statement is unfounded and unjustified. Indeed, it is contradicted by the excellent record of absence of failures of structures in service, although we are still learning about ensuring durability, contrary to the statement by Skalny *et al.* that 'we have known this (how to make durable concrete) for more than 50 years'. I have been 'in concrete' for more than 50 years and I well remember the days when durability was not mentioned explicitly in the codes.

Most designers and contractors are competent, and enforcement agencies are neither blind nor corrupt. Everyone is entitled to express views on a specific case but, being a structural engineer, I would be unwise to criticize chemists as a profession. Likewise, the writers of the letter, not being licensed and practising structural designers, should perhaps desist from wholesale criticism of the engineering profession.

To end on a more concordant note, I agree with Skalny *et al.* that 'the problem now is not lack of knowledge'.

6.5   GENDER IN CONCRETE

It is quite right to show no gender bias in our activities, including the language in our publications. There is, however, a practical point that I would like to raise. I hope there is no taboo on a discussion that some people might suspect of not being politically correct to such an extent that the topic must not even be aired.

First of all, I would like to establish my gender credentials, though I hope no one has doubts about my sex. When I was Head of the Department of Civil Engineering at the University of Leeds in the late 1960s and the 1970s, I strongly encouraged girls to study civil engineering, which had been considered the domain of males in wellingtons. We were highly successful in our efforts and we achieved more female students in civil engineering at Leeds than all the other civil engineering departments in the UK taken together. More importantly, our female graduates did very well, not only in a design office but also on a construction site. So I have no doubts that civil engineering, and its major component – concrete construction – is an activity for both sexes.

Indeed, I say so in the final paragraph of my book *Properties of Concrete* Fourth Edition, published in 1995. I quote: 'The first edition, and the subsequent two editions of this book, ended on a "tongue-in-cheek" note saying, "If the reader is unable to design a satisfactory mix he should seriously consider the alternative of construction in steel". The situation has changed. First, the reader may as well be a she as a he.'

There is no problem in languages in which gender and sex are decoupled. This is the case in German; for example, a girl is *das Mädchen* and a woman, or indeed a female, is *das Weib*, and both are neuter nouns. In French a person is *une personne* – a feminine noun – and that person, even if it is Napoleon, is referred to as *elle*.

Our problem with the written word arises from the fact that, in English, personal pronouns accord with the subject. There is no problem in French where, at least in simple sentence construction (no pun intended) the gender is not related to the subject; thus it is *l'ingénieur et son père* and *l'ingénieur et sa mère*, regardless of whether the engineer is a man or a woman.

On the other hand, in English, a male engineer's design is *his* work, and a female engineer's work is *her* work. To write *the engineer and his design* can be perceived to be gender-biased, not to say sexist. This, we are sometimes told, can be avoided by using the plural, like 'the contractor and *their* work'. But this is not logical in a situation where

266

there is only one contractor on a particular project. The alternative advice – nay, instruction – is to write 'he or she'. I find this cumbersome and distracting in reading, especially if used again and again. Anyway, does not 'he or she' give precedence to the male? It would be more fair to say, 'he or she, or she or he'. What a mouthful and how distracting!

Let me hasten to add that I am not advocating the use of 'he' on the understanding that 'he' subsumes 'she'. When I write papers, I try to construct my wording so as to avoid the need for these pronouns, but this is not always possible. What to do? To say 'the contractor and the contractor's programme' is inelegant and cumbersome. So, where am I going?

In legal cases in England, the contractor is referred to as 'it'. This is not insulting, but simply a recognition of the fact that the contractor is a legal entity, and not an individual. The same, generally, applies to the designer and the engineer. In the USA, the pronoun 'it' is sometimes used in articles written by lawyers. For example, in the March 2001 issue of *Concrete International*, Jeffrey W. Coleman wrote: 'A trial resulted regarding the issue of whether the engineer breached its contractual design responsibilities…'. Yet, when I write about 'a contractor and its performance', a politically correct editor edits it to read 'a contractor and his or her performance'.

There is an old saying arguing against gender discrimination: what is sauce for the goose is sauce for the gander; here, unusually, the saying starts with the female of the species. Perhaps we can argue that what is right for a lawyer is right for an engineer. We must not forget the difference between form and substance. 'He or she' is the form. It is the involvement of women in concrete construction that is the substance. And it is only the substance that matters, and it matters a great deal.

My plea is for a quiet re-think of the 'he or she' rule and consideration of alternative usage in the interests of clarity and simplicity, while strictly gender-neutral. I believe it is particularly appropriate to do so now that both the Concrete Society and the American Concrete Institute boast that they had a woman as President. Of course, I have no particular *right* to ask for a change. But we should not be too timorous to air the topic.

7

An overview

This is the last chapter in the book and, as such, it looks at changes that have taken place in concrete and concreting.

Section 7.1 reviews the changes in the last 40 years and questions whether all the changes represent progress, by which I mean significant improvement. Section 7.2 looks back at the last century; during such a time span there has undoubtedly been progress: compared with the early 1900s, our concrete is better. Specifically, we can prevent many types of attack on concrete, especially chemical in nature, and we can achieve much higher compressive strengths.

But we cannot take full advantage of these high strengths – up to 140 MPa or so – except in specific applications such as large bridges. Our codes of practice do not provide for design that routinely utilizes such high strengths. Why? An immediate answer is that we lack adequate experience to codify it. It should be remembered that codes of practice rarely provide for the use of novel and progressive methods, if only because codes rely on the collective experience of members of the code committee. This is not a criticism. For example, in the UK, prestressing was used for well over a decade before the publication of the first design code for prestressed concrete.

There are at least two other inherent reasons why very high compressive strengths are not utilized in design codes. One is that the increase in the compressive strength of concrete is not accompanied by a commensurate increase in its tensile strength and, after all, it is the behaviour of the tensile zone of a concrete member that often governs in flexure. A related limitation is the deflection of flexural members. This includes creep as well as the elastic deformation, and some of the positive influences on strength lead to an increase in creep deformation, which is generally undesirable.

Very high strength concrete is made with an extremely low water–cement ratio; the resulting concrete is almost impermeable, and this has an adverse effect on behaviour in fire: the water vapour formed at high temperatures cannot escape from the interior of a concrete member to the ambient atmosphere, and bursting and spalling may result.

So, in broad terms, there has been progress, but it was not spectacular. Basically the methods of mixing concrete, and consolidating (compacting) it have not changed. A man who last worked in a concreting gang in the 1950s (and there were no women on site then) would have no trouble fitting into a similar gang in 2006. Yes, robots are used to place and consolidate concrete, but almost exclusively in Japan. Yes, self-consolidating concrete is used here and there, but its use is not widespread.

What has changed is that we use many more admixtures, more fly ash, more ground granulated blastfurnace slag – but these are quantitative rather than qualitative changes. There is no leap-frogging in the concrete industry.

And yet, in many other fields the changes have been spectacular. When I was working on my first construction job in Taupo, New Zealand, in 1952, there was one public telephone in the village, and long-distance calls (say, 200 miles to the country's capital, Wellington) had to be booked and I was obliged to wait an hour or two, or even find that the post office (where the sole public telephone was located) was closed for the day or, worse still, for the weekend. The reader does not need to be told that, nowadays, every 5-year old has *its* mobile telephone, probably equipped with some kind of video imaging.

On the somewhat larger scene than concrete as a material, Section 7.2 notes that there has been some unification of design codes. In Europe, uniform design codes are being developed, and standards for cement have been made identical in all countries. This is no mean achievement, given that code writers, who inevitably have many years' experience with their own national approach, must consciously depart from tradition. And this on a continent with a long history of wars and rather strong nationalistic attitudes.

Had I been writing this book, say, 6 years ago, I would have said that we were all moving toward using the SI system of units of measurement. The word 'all' must include the USA – the largest single economy in the world. Alas, in recent years, the USA has retrogressed: the individual

269

States no longer require the use of SI units. Even ACI – the prime concrete organization in the world – has reversed its policy of using SI units as primary values and dimensions. I cannot help smiling when I think that the old Imperial bonds of feet, square yards, and pints still hold strong: the British colonization was, after all, a durable success.

Although a heading in Section 7.3 includes the word future, I have refrained from daydreaming and wishful thinking. The traditional *forte* of concrete is that it can be produced almost anywhere on earth, using local ingredients. No other material threatens to match this situation. The second strong point of concrete is that it allows the use of an unlimited choice of shapes. To date, this has been under-utilized, and we should encourage architects to depart from strict rectilinearity.

The penultimate section (7.4) looks to the future. It is not a prediction. Rather, I express my views on what needs – nay, must – be done for concrete to continue to have a pre-eminent position in construction. Not only do I discuss the necessary improvements in field practice, but I also postulate a dichotomy. The first type should be ordinary concrete 'for everyday use' relying on cement properties 'as delivered' and on simple, relatively cheap mixes. The second type of concrete should be tailored to a specific use that is fit for the purpose. This material would be much more expensive because of the specific selection of Portland cement and other cementitious materials, as well as of admixtures, but would be durable and serviceable with a minimum of maintenance and repairs.

The present situation in the UK, where more than 40% of the total expenditure on construction is devoted to repairs and maintenance is unsatisfactory. Maintenance is well worthwhile, but many repairs could be avoided by better quality initial construction.

Finally, Section 7.5 is a personal farewell note. I have enjoyed my very long association with concrete and concretors, although it was not quite so long or quite so pleasurable as my 75 years of skiing, which came to an end two years ago.

Now it is good-bye to concrete. This is my tenth and last book. I would like to quote Hilaire Belloc who wrote:

> When I am dead, I hope it may be said:
> 'His sins were scarlet, but his books were read'.

7.1   CONCRETE: 40 YEARS OF PROGRESS??

I should start by explaining the title of this section. The question mark is to indicate that I do not necessarily view the changes that have taken place as representing significant progress. However, the second question mark is there to show that my text is meant to be provocative rather than a simple catalogue of changes in the last 40 years.

I should also explain why I have chosen an interval of 40 years. The main reason is that, by 1963, most of the scientific work that enables us to understand the basic properties of concrete was established: permeability, porosity, hydration and development of strength, some of the mechanisms of chemical attack, and resistance to repeated freezing and thawing. We had a basic knowledge of shrinkage, creep, and temperature effects.

An additional reason for my choice of 1963 as the starting point is that, in that year, the first edition of my book *Properties of Concrete* was published. From that edition onwards, I can trace the changes through the second, third, and fourth editions of that book up to 1995; the 12th impression of that edition appeared in 2005. Further changes and the practical problems of recent years are discussed in *Neville on Concrete*, published in 2003.

Portland cement

The five fundamental types of Portland cement, ordinary, rapid-hardening, low-heat, sulfate-resisting, and (at least in the USA) modified (or moderate) cement were classified and established by 1963. They have served us well ever since, but occasionally problems arise when users fail to realize that the types are not mutually exclusive: a given cement can satisfy simultaneously the requirements of being ordinary and rapid-hardening, or ordinary and low-heat, or low-heat and sulfate-resisting, or ordinary and sulfate-resisting. The explanation lies in the fact that the specifications for the various types of cement prescribe minima or maxima for some physical properties and for the amounts of the various chemical compounds, which, in any case, are no more than calculated potential amounts.

Although the fundamental types of Portland cement have not changed in 40 years, there have been significant changes in certain properties of cement. Most of these changes stem from the new methods of manufacture of cement. These methods are more economical in energy consumption and in the production of cement, and they also allow a

271

wider range of fuels to be fed into the kiln: powdered coal, oil, gas, petroleum coke, biomass, and old tyres. Some of these fuels introduce more sulfates; these sulfates are more or less soluble, and are therefore available for reaction within the hydrated cement. That solubility is also relevant to the action of superplasticizers, which I shall discuss later on.

Other changes in the properties of Portland cement arise from the retention of alkalis in the cement and from grinding to a finer powder. There have also been changes in chemical composition. Broadly speaking, ordinary Portland cement manufactured in the last 25 years gains strength more rapidly than 40 years ago. There are commercial benefits to this: formwork can be struck earlier and construction can proceed more rapidly. However, this results in a shorter period of effective curing.

Why then is it that improvements in the manufacture of cement have led to problems in structures? For years, the primary criterion of acceptability of concrete was its 28-day compressive strength. With the 'new' cements, the same strength as before was achieved at a higher-water–cement ratio, so the contractor was justified in using this ratio, and therefore a lower cement content, for the same workability as before. The resulting concrete had a higher permeability than the concrete made with the old cement and was therefore more liable to carbonation and to penetration by aggressive agents.

The problems arose because designers did not appreciate the fact that, although the specification remained the same, the concrete in the structure was inferior in long-term quality, especially with respect to durability.

Cementitious materials

There have been important changes in the composition of what I shall call cementitious materials; by this I mean the very fine inorganic powders incorporated in the concrete mix that have hydraulic properties or potentially hydraulic properties, and also those that are chemically inert but have physical characteristics contributing to the strength-gaining properties of cement.

In the past, a material satisfying British Standards for cement had to be pure: that is, it was to be made exclusively from cement clinker, gypsum, and grinding aids. Nowadays, the European Standards, which of course are British Standards as well, allow routinely a proportion of an inert filler.

Moreover, European Standards offer a wide range of cementitious materials. Some of these were known 40 years ago, especially ground granulated blastfurnace slag (or slag, for short) and pozzolanas, more often natural than artificial. Generally, they were used where available locally and only as a fairly limited proportion of the total cementitious material. Today, the use of either slag or pozzolanas, and also both in the same mix, is widespread. A particularly popular pozzolana is siliceous fly ash, also called pulverized-fuel ash. This is the ash precipitated electrostatically or mechanically from the exhaust gases of coal-fired power stations.

Fly ash is described nowadays as a by-product, but 40 years ago it was considered to be a waste product. I remember clearly that coal-fired power stations offered the fly ash for free, provided a lorry was sent to collect it. The problem was that fly ash occupies a large volume and is not compactible so that its disposal, or transport for that matter, was not easy.

Fly ash is no longer cheap but it bestows various beneficial properties on concrete: the rate of hydration of the cementitious material containing fly ash is slow so that thermal cracking is readily avoided; the fly ash removes calcium hydroxide from the products of hydration of Portland cement, thus making it less vulnerable to some forms of chemical attack. Comparisons of strength of mixes with and without fly ash are not very meaningful because these mixes do not contain the same amount of Portland cement, and they may or may not have the same total content of cementitious material or the same water–cementitious material ratio. Moreover, the rate of development of strength of mixes containing fly ash is slower than with Portland cement alone, which may or may not matter.

The only *practical* basis for comparisons is that of cost, but the price of fly ash varies from place to place; in particular, the price is affected by handling and transport costs from the source of fly ash or from its vendor: that is, the power generating plant or the cement manufacturer who may control the sales, especially when the cement and fly ash are blended.

What I would like to emphasize is the beneficial effects of fly ash, which is ever more widely used and also used as a greater and greater proportion of cementitious material, sometimes in excess of 50%.

Other siliceous materials – and the silica must be in amorphous form – are used in some parts of the world, for example rice husks and metakaolin. Metakaolin is expensive, and rice husks, which have to be fired at 500 to 700 °C, have a high water demand unless interground with clinker so as to break down the porous structure due to the plant origins of the husks. Intergrinding means that clinker and burnt husks have to be sent to the same location, this necessitating transport, and rice husks are plentiful in countries in which transport is often not easy or cheap. They are also available at four locations in the USA. If the husks are not interground, superplasticizers need to be used, and they are expensive: this negates the cheapness of rice husks. Anyway, I have to admit that I have no experience of rice husks, but their use is limited to countries where rice is the staple diet.

From the 40-year perspective, I have to applaud the shift from using solely Portland cement in concrete to the use of a range of cementitious materials, with Portland cement being only one component, albeit an essential one. Here, I would like to drop a bombshell – something I have done also in the past. Portland cement is a necessary component of concrete, but it is also the source of many problems in, and with, concrete. Let me give a few examples.

Perhaps the most common problem with concrete is shrinkage and the consequent cracking. Now, shrinkage takes place within the hydrated cement paste, so that it is cement that is responsible; after all, with very few exceptions, aggregate does not shrink. Creep of concrete is not always harmful, but usually it has undesirable effects, notably a time-dependent increase in deflection and also a loss of prestress, and it is the hydrated cement paste that undergoes creep. The development of the heat of hydration and the consequent expansion and contraction of concrete, with the possibility of thermal cracking, are due to the hydration reactions of Portland cement.

There are other deleterious consequences of the presence of Portland cement in concrete: the alkali–silica and alkali–carbonate reactions are induced by the alkalis in the cement (unless low-alkali cement is used); sulfate attack involves tricalcium aluminate in the cement. Some types of chemical attack on concrete involve leaching of calcium hydroxide, which is a major product of the hydration of cement, or even of calcium silicate hydrates, which originate from the same source. In some climates, repeated freezing and thawing leads to a disruption of hydrated cement paste, unless precautionary measures, such as air entrainment or an extremely low water–cement ratio, are used.

Let me emphasize, we cannot make concrete, as we know it, without Portland cement, but we should minimize the cement content, balancing technical advantages and disadvantages on the one hand, and on the other, cost. In the last 40 years, we have developed and improved the means of doing so: other cementitious materials, which I have already discussed, and chemical admixtures.

274

Admixtures

The last 40 years have seen a great increase in the use of admixtures. The USA has always been in the forefront but, in the UK, the main admixtures used in 1963 were water-reducing admixtures, some retarders, and calcium chloride as an accelerator. Calcium chloride contributed to the corrosion of reinforcement, and it took until 1977 to ban it.

Not surprisingly, British cement manufacturers were not enamoured of admixtures. I remember becoming enthused about admixtures after a visit to the USA and I was 'preaching' about their benefits when I was President of the Concrete Society in 1974–1975, only to be told by a cement manufacturer: the best admixture is more cement. Of course, it is not! Attitudes have changed since, and today some cement manufacturers also sell admixtures. The slow and cautious British approach gave rise to the story about an American who was asked where he would like to be when the world comes to an end. In England, he replied: everything there arrives 10 years later.

The greatest benefits occur when superplasticizers are used. They allow the use of extremely low water–cement ratios, even below 0.30, with adequate workability. They are expensive, and their compatibility with a given Portland cement has to be ensured, which requires testing on a case-by-case basis. One of the reasons for this is the variable reactivity of SO_3 in the cement clinker, and this variability is the consequence of the type of fuel used to fire the kiln. I am convinced that new and better superplasticizers will increase their use; possibly matched pairs of cement and superplasticizer will become commercially available.

Aggregate

As is well known, some three-quarters of the volume of concrete is occupied by aggregate. Moreover, it is essential that the greater part of the volume consists of aggregate. This is so not just because, on a mass or volume basis, aggregate is cheaper than cement, but also because of the disadvantages inherent in hydrated cement paste, discussed in an earlier section. Above all, a neat cement mix would be difficult to compact, except at a unique water–cement ratio. The neat cement would generate enough heat to boil an egg, and then would cool, contract and crack.

Aggregate is important. Now, what has changed in the last 40 years? By 1963, we had a good knowledge regarding various aggregate types in

275

relation to density, porosity, strength, abrasion properties, crushing strength, moisture content, clay content, unsound particles, organic impurities, thermal properties and salt contamination. We had a number of rules of thumb about desirable gradings, which are incorporated into British and American Standards. In particular, the Road Research Laboratory in the UK produced a great deal of practical information.

Much of our knowledge of the influence of aggregate grading on workability and on compaction referred to natural aggregates, whose size, shape, and texture were the result of weathering. However, in the last 30 years, in many parts of the world, such aggregates have ceased to be available, or winning those aggregates from riverbeds and pits has become unacceptable on environmental grounds, if not outright prohibited. We have been forced to use crushed rock, and the process of crushing produces particles with surface texture and shape properties different from natural aggregate. Alas, we have developed no detailed methods of describing these properties in a manner that can be synthesized to provide a description of the bulk of the aggregate to be included in the mix. Also, we know very little about recycled aggregate from demolition and from processed waste, but we must use these materials for ecological and economic reasons.

We have continued to rely on grading curves derived from a series of sieves whose size decreases in geometric progression, such that the opening of any sieve is approximately one-half of the opening of the next larger sieve size. Metrication introduced some complications, which were largely resolved by the European Standards, leading to sizes in mm of: 63.0, 31.5, 16.0, 8.00, 4.00, 2.00, 1.00, 0.500, 0.250, 0.125, and 0.063. This is still the situation today.

The real problem lies in the fact that if, say, 20% of a given aggregate is smaller than 2.00 mm or just larger than 1.00 mm, much of this 20% could be *just* smaller than 2.00 or *just* larger than 1.00 mm. To overcome this situation, we have used grading zones, and proscribed gradings that meander between zones. But in real life, there is an infinite number of gradings all of which have the same percentage passing *each* of the standard sieve sizes, but which differ greatly in their behaviour with respect to workability or water demand. Nothing has changed in this respect in the last 40 years.

Likewise, nothing has changed to make it possible to provide a mathematical description of shape or texture of aggregate particles, both of which greatly influence the workability of mixes with nominally similar gradings. And yet, physicists are able to measure, describe, and prepare surfaces in a highly specific way, not to mention the advent of nanotechnology. I think we ought to be ashamed of this lack of progress. Ashamed, yes, but not surprised. I am saying this because the relevant research would have to be paid for, and the paymaster must reap a benefit from his investment: the benefit would be for *all* users, and not for a single enterprise. Furthermore, the cost of producing aggregate with specified surface properties would increase greatly, and this is not in anyone's apparent interest. I shall explain the term 'apparent' in a later section.

Water

Water is an essential ingredient of concrete necessary to produce a workable mix and to allow the hydration of cement to take place. So we ought to know something about water.

First, let us look at the quality of mix water. Generally, there is no problem there because we specify the use of tap water and, nearly always, such water is satisfactory for mixing and curing purposes. But not all concrete is placed within reach of a city water supply, and much of the modern construction takes place in arid areas where the supply of water is often problematic.

A decision on the suitability of a given water should then be based on its analysis. We know that the water should not contain oils or fats, excessive humic matter or solids in suspension. As for the chemical compounds, no systematic testing has been done during the last 40 years, so that we have to rely on a rule-of-thumb approach. The best set of such rules is provided in a European Standard published in 2002. This standard is probably conservative in the limits imposed so that we need not fear problems in concrete. However, erring on the safe side may result in wrongly rejecting some waters, and in importing more expensive water from a source that conforms to a standard.

As far as the American practice is concerned, in the 1963 edition of *Properties of Concrete*, I relied on an ASTM publication dated 1956. In the Fourth Edition of *Properties*, the relevant ASTM publication was said to have had 'minimal updating'.

Why has no systematic research been conducted to determine safe limits of sulfates, nitrates, aluminium, and lead or zinc? Such a research project would require extensive testing and would be fairly expensive. Who is to pay for it? The paymaster would need to receive a benefit, and no benefit would accrue to water companies or to cement manufacturers. So, there is no apparent interest, and no studies have been conducted.

The quality of water to be used for curing purposes is also important, but the requirements for curing water and for mix water are not necessarily coincident. For example, water containing excessive solid matter or alkali carbonates is satisfactory for curing, although not as mix water. Likewise, wash water from a concrete plant containing alkali carbonates or a residue of admixtures is harmless when used for curing, but it is not suitable as mix water. Seawater is certainly not suitable for curing when the concrete is very young because imbibition of chlorides may lead to corrosion of reinforcement. I have not seen any *general* guidance on this developed in the last 40 years.

Workability and compaction

First, how do we measure workability? Generally by slump, a test method more than 80 years old, and a cruder method it is difficult to imagine. Cohesion of the mix and absence of segregation are vital, but we judge these by eye. Admittedly, numerous gadgets and techniques have been developed over the years, but none of them has gained acceptance because none measures any fundamental or specific property of the mix in the *field*. They all work, and so does the slump test, as long as all we want to do is to detect local and periodic departure from a given mix during an ongoing construction.

Water in the mixer comes from a tank in a batcher and from water adsorbed on the surface of the aggregate. We determine the adsorbed water by testing, usually quite crudely, samples of aggregate from the stockpile. But this moisture content, even in a well-drained stockpile, varies vertically, horizontally, and varies also with time. So the input of this water into the mixer varies, sometimes widely, and we resort to corrections by the mixer operator on the basis of his observation of the mix or from measurement of the torque necessary to rotate the mixer. But the torque is affected also by the grading of the aggregate, so that the whole technique is very crude.

In consequence, last minute adjustments to the quantity of water are sometime made at the point of delivery of the concrete into its final location. This is done by adding water into the agitator drum, sometimes surreptitiously, and a quick re-mix. I am not criticizing such procedures because, in order to be compacted by the means available, the mix *must* have the desired workability. This is even more important than adhering to the specified water–cement ratio because a mix that cannot be adequately compacted is bound to be highly unsatisfactory.

Nevertheless, not exceeding a specified maximum value of the water–cement ratio is of great importance but specifications often prescribe a minimum cement content. Alas, the requirement for a minimum cement content continues to be heavily embedded in specifications. And yet, there is no scientific basis for the belief that more cement makes better concrete. All that a high cement content does is to ensure a sufficiently low water–cement ratio at an acceptable workability; both of these should be controlled direct at the mixer.

Now, what do we know about adequate consolidation (compaction)? It is easy to describe it in terms if achieving the theoretical density of the mix, but we cannot measure that in situ. We know what we want, but in the last 40 years no suitable technique has been developed to enable us to verify the degree of compaction in situ, other than by taking cores, which is too late.

Curing

This final procedure in making concrete is a Cinderella of construction. Curing is not paid for as a separate item in the bill of quantities, and rarely is there a curing specialist on site. This situation has hardly changed in the last 40 years. And yet, if anything, the need for curing has increased because, as I have already mentioned, more construction takes place in arid zones than 40 years ago.

I am not denying significant developments in membrane curing, where product manufacturers have something to sell. Opinions differ on the efficacy of membrane curing, but it is not really suitable for concretes with a water–cement ratio of about 0.35 or less. Likewise, there are frequent difficulties with wet curing of vertical surfaces, especially when slipforming is used.

All in all, I think it is a pity that curing is not thought to be of great importance, and I have seen many cases of well-designed and well-placed concrete that deteriorates early in its life because of the absence of wet curing.

Areas where progress has occurred

The great increase in the use of ready-mixed concrete and the extensive use of pumped concrete represent a considerable improvement in concrete practice. So does the use of superplasticizers, which permit placing mixes with extremely low water–cement ratios. Nevertheless, because of their cost, superplasticizers have not become a run-of-the-mill commodity.

Another significant change is the move from prescriptive specifications to a performance-type approach. This gives more scope for benefiting from a knowledgeable and efficient ready-mixed concrete supplier. If good profit can be made by supplying satisfactory concrete at a lower price, then this is good for all concerned.

What is important is that the variability of concrete as placed should be minimized. We sometimes forget that mix proportioning is aimed at ensuring that 99% of the mix as tested exceeds the given property, for example the design strength, but the cost is related to the mix corresponding to the value of the property, usually strength, that is satisfied by 50% of the tests. A simple example is the requirement for the design strength of 40 MPa; if the standard deviation is 3.5 MPa, then the mean strength is $40 + 3.5 \times 2.33 = 48$ MPa. The actual situation is complicated by the test method used: specifically, whether a test result is a single cube or cylinder, or the average of two or three specimens tested at the same time. The important practical point is that the cost of production is based on a 48 MPa mix but all the designer really needs is a 40 MPa mix. The spread between the design strength and the mean strength can be reduced by a better batching operation. Much progress has been made in this respect in modern batching plants.

Why has progress been limited?

Earlier on, I referred to the lack of an economic incentive for research on topics such as the quality of mix water, or aggregate properties at the batching plant. Yet, much research continues to be undertaken at universities. Alas, much of it is purely scientifically oriented, rather than directed towards making better concrete in ordinary construction.

Thus, we now know a great deal about fibre-reinforced concrete or glass-reinforced cement, but these are highly specialized materials. University researchers have also written literally many thousands of papers on alkali–aggregate reaction, on delayed ettringite formation and, more recently, on thaumasite sulfate attack. I see such work as a disproportionate effort from the standpoint of the popular need for good concrete.

Concluding remarks

The answer to the question posed in the title of this section is: yes, we have made progress over the past 40 years. But not as much as in other technical fields, such as electronics or computing, or computer-controlled operations. Car manufacture relies significantly on robots but, outside Japan, the use of robots in concrete is all but unknown.

Admittedly, the construction industry as a whole is still highly labour-intensive, with concreting bringing up the rear. The time has come to leap forward. I believe that, to achieve this, we need not new scientific research, but the application of scientific knowledge already available and of techniques from other fields. If concrete is to keep its pre-eminent position in construction, now is the time to apply the available scientific knowledge about cement and concrete, as well as techniques used in other fields, to the concrete industry.

Bibliography

Neville, A.M., *Properties of Concrete*, Fourth Edition, 1995, 12th impression with standards updated to 2002, Pearson Education, Harlow, 2005.

Neville, A., *Neville on Concrete*, ACI, Farmington Hills, MI, 2003.

7.2 LOOKING BACK ON CONCRETE IN THE LAST CENTURY

In 2004, the American Concrete Institute (ACI) celebrated a century of its existence. Strictly speaking, ACI was originally established as the National Association of Cement Users, which in 1913 became ACI. At the celebrations in Washington, DC, I was asked to address the International Luncheon, and I chose as my title: '100 years of ACI as seen by an 80-year-old non-American'. Given that Americans generally are inclined not to take great interest in the world outside the USA, I felt it useful to point out that I lived on a small island called Great Britain and that we used to run the world before the Americans. The long-term historical perspective is best illustrated by a cartoon in the *New York Herald Tribune*, in which one American reflects on the days of taxation without representation, which ultimately led to the Boston tea party, a rebellion, and independence from Great Britain. The second American replied, 'That three-per-cent tea tax don't look so bad nowadays.'

Origins of concrete organizations

My address made me think about an 'anniversary' article. And there are other anniversaries too, although not all of them are nice round numbers. At home, the progenitor of the Institution of Structural Engineers was established in 1908 as The Concrete Institute and in 1923 adopted its current name.

Our own Concrete Society was established in 1966 and so is 40 years old, and it is still forging ahead.

As for myself, I have spent more than half-a-century in concrete – well, on and off – and the time has come to hang up my boots; in my case they are gumboots. All this is a somewhat spurious justification for writing a sort of farewell section, which looks back on concrete, mainly in my lifetime.

Changes in concrete

I suppose the starting question should be: has much changed? The answer is in two parts. As far as cement is concerned, there have been very significant changes. In so far as concrete and concreting are concerned, the changes have been much more modest.

I am sure all the readers of this book are crystal clear about the difference between cement and concrete, but the two are often used interchangeably, even by people who ought to know better, and certainly by the popular press. The way to drive the distinction home is to say: cement is to concrete what flour is to fruitcake.

My reference to hanging up my gumboots was not just a corny expression; what I wanted to emphasize was that efforts to improve concrete structures through better concrete require research on practical concrete and on concrete practice. This is what I have attempted to do in the last 10 years or more: *vide* my book *Neville on Concrete* whose subtitle is *An Examination of Issues in Concrete Practice*. This book carries my effort further. Of course, research on laboratory concrete is also required and so is research on cement. The latter should be directed towards obtaining better cements or cements particularly suitable for various purposes. What I deplore is research on cement, which purports to answer questions about concrete by simply ignoring the properties of the aggregate or, above all, the transition zone between the aggregate and the cement paste.

I suppose one reason for the above situation is that many researchers are not civil engineers and have no understanding of, or interest in, structural engineering. They are usually chemists, sometimes chemical engineers, often materials science graduates.

I must not go overboard. The scientific research about or into concrete is desirable, nay necessary. But the position of academic civil engineers who are not civil engineers by training should not exceed, say, 20 to 25% of staff members. I am convinced that in many departments it is much higher. Whether the holder of a PhD degree who has never worked in a design office *and* on site is best qualified to train budding civil engineers is a question that merits serious consideration. All I can say is that, when I was responsible for appointing lecturers in civil engineering, I set much greater store upon an applicant's membership of the Institution of Civil or Structural Engineers than upon having a doctorate.

Internationalization of standards

Let me return to ACI: I think it has a very special position in the world of concrete because it is truly an international body. This does not detract from our own Concrete Society, of which I was proud to be the President in 1974–1975. We continue to need the Society, as well as all the other 'concrete bodies', now grouped in one physical location.

However, in certain respects, the world is changing and becoming more internationalized. We are already seeing the British Standards being replaced by European Standards, and this trend will continue until the only remaining British Standard will be that for haggis, and that will appear under the aegis of the Scottish Executive! The design codes have also been, or are being, Europeanized. Putting national pride aside, there is no rational reason for loads, or actions, deformations or strains, and methods of analysis to differ from one European country to another. The next generation of designers will not even know that each country had a separate design code, just as few of the current designers are aware (as I am) that once upon a time there was a code issued by the London County Council.

If the logic of moving from British codes and standards to European ones is accepted, as it must be, then we can extend it to a transatlantic expansion. By this, I certainly do not mean that we simply subscribe to ACI codes and adopt them as our own. The process of amalgamation should be achieved by discussion, coordination, adaptation, and compromise.

Once this has been done, we would reap some advantages. For example, there would be no barrier to British designers working on projects in America – and not just North America, because many South American countries also use the ACI Building Code, ACI 318.

Furthermore, it would be easier for us to use American Codes where a British code does not exist. A few years ago, I was involved in a major dispute concerning the collapse of a circular cement silo consisting of nine compartments. There is no British code for that type of structure; the only guide extant is a guide published by the British Materials Handling Board in 1987, but I do not think it has a recognized standing. On the other hand, there is an ACI Code, ACI 313-97 'Standard practice for design and construction of concrete silos and stacking tubes for storing granular materials'.

Now, it is well known that designing a structure by picking isolated parts from different codes is unwise, if not outright dangerous, because inconsistencies and incompatibilities may lead to spurious results. If the general, underlying code had been from the same source as the silo code, there would have been no problem in checking the original design on the basis of the silo code. For that mater, both codes could have been used in the original design of the silo.

As far as standards for materials and test methods are concerned, the American Society for Testing and Materials (ASTM) has an excellent and very large portfolio of standards. This covers some fields for which no British Standard exists, and some ASTM standards are used in this country. Again, there is scope for adopting common standards with the USA. Incidentally, ASTM is now called ASTM International.

I am aware of some of the problems in preparing European standards that arose from different technical and cultural backgrounds in the various countries, not to mention language problems. I can give an example from the days when I was Chairman of the RILEM Permanent Commission on Concrete, which prepared a number of standards for test methods; these had to appear simultaneously in English and French. The language problem can be illustrated by the following.

In a British standard for a test, expressions such as 'should not be tested' or 'shall be made as thin as possible' are used. On the other hand, in French, the imperative is used unashamedly: do not test. It worries me that lawyers or courts might interpret the English and French versions differently.

So, there would be difficulties but they would not be insuperable. With modern electronic communication, virtual meetings are easy and there should be no delay in arriving at a solution.

Units of measurement

Having said all that, I am aware of the fact that the USA and the rest of the world are separated by a fundamental difference in the units of measurement. This is not just a question of numbers but also of a feel for numbers.

In my own experience, my engineering education was in the old British system of pounds and feet, including pound force, pound mass, and even slug (which is numerically larger than a pound by the gravitational constant, but the concepts of the two differ). At the end of the 1960s, like everybody else, I had to convert to the SI system, and eventually I forgot all about the old units of measurement. Some 14 years ago, I found myself acting as an expert witness in law courts in the USA in cases involving – surprise, surprise – concrete construction. While I had no difficulty in a rapid mental conversion of feet into millimetres or pounds per square inch into megapascals, when it came to pounds per cubic yard I had no 'feel' for the values involved.

The fact is that the Imperial system, to which the Americans cling in a surprising manner, is irrevocably doomed. I know that the USA economy is so large that the Americans think they can ignore the rest of the world. I submit that they cannot, or cannot for long. The American powers-that-be are aware of that situation. In 1975, the Metric Conversion Act designated the SI system as the preferred system of measurement.

In 1988, the Omnibus Trade and Competitiveness Act mandated SI as the preferred system of weights and measures for trade and commerce.

However, the concrete industry and the ACI showed themselves to be less progressive, probably because concrete generally does not travel across borders, even though many American Federal agencies introduced the SI units into the design of all federal construction. As far as ACI codes and guides are concerned, some of the most important ones, notably the Building Design Code ACI 318 exist also in the SI form. Recently, however, ACI has slow-pedalled on metrication, and in the spring of 2004 it decided that 'all new and revised ACI specifications, standards, and other publications shall use dual units'. This is hardly likely to lead to a conversion to SI units; on the contrary, the Imperial system has been given a new lease of life.

In a way, the above is ACI's own business but, as H.S. Lew, Chairman of the ACI International Committee, wrote in 1993: 'If ACI truly is an international organization, shouldn't it be communicating in a measurement language officially adopted by most countries in the world?' What concerns me is that when, in 2004, I wrote: 'internationalization of construction is hampered by the way some people in the USA cling to the old units', these words were censored by an apparatchik.

Censorship of technical publications

The notion of censorship of technical publications worries me. I am not referring to general political correctness, although this sometimes forces us to write ungrammatical English, such as 'a particular contractor *was* obliged to withdraw *their* bid'. Avoiding singular pronouns because some people see sexual connotations in them is, to my mind, exaggerated.

Likewise, when I saw my quotation from St Luke 'The labourer is worthy of his hire' changed by the editor of a journal to 'The labourer is worthy of his or her hire', I felt that readers would think me peculiar, to say the least.

Censorship that bears on technical knowledge in practice is much more serious. A few years ago, I looked into the leaching of some chemical compounds and elements from concrete pipe or mortar-lined steel pipe carrying drinking water, or more generally, water intended for human consumption. This leaching takes place when water is stationary, or mainly so, which occurs during the night.

The attack by hardened cement paste may raise the pH of the water and increase its carbonate alkalinity or water hardness, induced by the carbon dioxide dissolved in the water, which reacts with calcium hydroxide in the cement paste. This could be described paradoxically as attack *on* water *by* concrete. Leaching by the water may also result in an increased content in the water of aluminium, calcium, sodium, and potassium. Some corrosion inhibiting admixtures used in the concrete mix may also be leached and present a danger to health. Likewise, some curing compounds and formwork release agents may persist on the surface of the concrete and eventually enter the water.

The present section is not a forum for a detailed technical discussion; suffice it to say that the problems are real and serious enough to have led to some European directives, and a European Approval Scheme for water conduits is expected in the near future. And yet, when I submitted a serious paper on this topic, I was very politely advised to seek another venue for publication because the pipe manufacturers were considerable contributors of funds to the institute publishing the journal.

I had no problem publishing my paper in another highly reputable journal, but I am worried that, if an independent, or seemingly so, concrete organization protects its financial sponsors at all cost (no pun intended) then the world at large is a loser. And if, in the fullness of time, there are technical problems because no warning was issued, then the good name of concrete will suffer, and this will impinge on all of us in concrete. The Duke of Wellington was right when he said, publish and be damned.

I am discussing this censorship, self-imposed by a highly reputable concrete organization, at length because a year ago, ACI put a taboo on the discussion of sulfate attack on concrete because this topic is the subject of numerous lawsuits in California. Clearly, no individual dispute should be the subject of an article without agreement of the parties involved, but a general, cool, objective, and scientifically based discussion is essential if we are to learn from our mistakes and do better in the future. And it is precisely this that is, or should be, the objective of much writing on concrete. Publishing nothing but glossy pictures of successes may be pleasing but it does not lead to progress.

Investigations of problems

The preceding lengthy discussion of censorship leads me to the consideration of investigations of problems in buildings. This was the topic of an excellent paper by Brian Clancy in the *Structural Engineer* (6 January

2004). He encapsulated the didactic value of such investigations in the words: 'You will learn more about buildings and structural engineering from sorting out problems than you will ever learn in the office doing new-build design.'

The relevance of investigations of problems and failures is that reports on these are of widespread interest, and yet some journals have a ban on the topic for fear of offending someone, possibly influential. Clearly, a publication on these delicate topics requires permission of the party that employed the writer, or else it has to be couched in such terms that the project cannot be identified. With these safeguards, there should be no problem with people of good will. However, never to publish a paper on, say, sulfate attack on concrete (as is the ACI policy) is to pretend that there is no attack or no problems consequent on that attack. This cannot be conducive to progress.

7.3   CONCRETE PAST AND PRESENT

The starting question should be: has much changed? In so far as concrete and concreting are concerned, the changes have been much more modest than in, say, electronics or communication in general.

Efforts to improve concrete structures through better concrete require research on practical concrete and on concrete practice. This is what I have attempted to do in this book and previously in numerous papers and articles published as a collection in my book *Neville on Concrete* whose subtitle is *An Examination of Issues in Concrete Practice*. Of course, research on laboratory concrete is also required and so is research on cement. The latter should be directed towards obtaining better cements or cements particularly suitable for various purposes. What I deplore is research on cement, which purports to answer questions about concrete by simply ignoring the properties of the aggregate or, above all, the transition zone between the aggregate and the cement paste.

I suppose one reason for the above situation is that many researchers are not civil engineers and have no understanding of, or interest in, structural engineering. They are usually scientists, working in a laboratory and more interested in the underlying science than in the 'how' and 'why'.

Great achievements

To restore the balance of criticism, I have to record the most impressive series of landmark papers, re-published by ACI in 1994. These are truly seminal papers originally published in the last century, dealing with fundamentals of the physics of cement paste (T.C. Powers), practice of concreting (Ben C. Gerwick, Ed Abdun-Nur), structural analysis and design (Hardy Cross, Douglas McHenry), durability (Paul Klieger), and other topics.

One of the other topics was entitled 'Small concrete houses at Rochester, N.Y.', and was published as far back as 1922. The author was Kate Gleason, a woman contractor. She was very successful financially, but also technically in that she introduced concrete houses in a country where timber is, to this day, a pre-eminent material in housing. It is people like this that should be remembered.

The introduction of concrete houses was not easy. The *Structural Engineer* of 2 December 2003 refers to 'a well known advertisement where a very large block of concrete was cast and then hollowed out by several men using hammers and chisels'. This is hardly an example of buildability, and I presume it was meant humorously.

A real difficulty was encountered in Norway, many years ago, where timber construction was the norm: a concrete enthusiast tried to persuade an architect to design concrete houses. The 'concrete man' painstakingly described the use of timber formwork to cast walls, including provision for apertures. When he finished, the architect said: 'Let me check whether I have understood you correctly. You build a timber house, then an outer timber house. You fill the space in-between with concrete. Then you throw away the outer timber house and the inner timber house, and so end up with a concrete house, which is cheaper than a timber house.' The advent of concrete housing was slow.

It is only fair to record that Thomas Alva Edison, best known for his invention of the incandescent lamp and of the phonograph, contributed also to cement production, and around 1909 wrote: 'I believe I can prove that the most beautiful houses that our architects can conceive can be cast in one operation in iron forms for a cost which will be surprisingly low.' Note the early use of metal forms.

Past progress

It would not be appropriate in this short section to review in detail the various developments in concrete. At the risk of offending some people, I would say this. The middle of the last century was glorious. First, there was the fundamental work on the physical properties of hydrated cement paste and on the chemistry of Portland cement, much of it done at the Portland Cement Association in the USA. At the same time, there was seminal work on aggregate properties and grading and on compaction, leading to a rational approach to mix design, done at the Road Research Laboratory in the UK. The significant British contribution to the chemistry of cement and concrete achieved at the Building Research Station culminated in the book *The Chemistry of Cement and Concrete* by Sir Frederick Lea in 1935 and 1956.

The British contributions are not always recognized, even in the UK. At a recent presentation to the Parliamentary and Scientific Committee, an eminent Oxford Professor of Materials Science listed a dozen or so great British inventions and innovations from the steam engine to stainless steel. I was obliged to point out the omission of Portland cement, patented in 1824 by Joseph Aspdin, a Leeds builder.

In the last quarter of the 20th century, there was a slow-down in progress. An important development was the much greater use of admixtures in the UK, which had already been widely used in the USA for a number of years. I understand that in 1975, 12% of concrete placed in the UK contained admixtures compared with 70% in the USA, and 80% in Australia and in Japan.

I remember the adage of some British cement manufacturers to the effect that the best admixture is more Portland cement. This is definitely not so: while you cannot make ordinary concrete without Portland cement, that cement is responsible for all the main problems in concrete. As discussed in Section 7.1, the problems whose seat is undoubtedly the hydrated cement are: shrinkage and consequent cracking; creep and the associated increase in deflection; excessive heat development leading to cracking; damage caused by freezing and thawing; alkali–aggregate reaction; external chemical attack; and carbonation leading to steel corrosion.

The remedy or, more correctly, an alleviation of the problems was achieved by increasing the proportion of other cementitious materials at the expense of a reduction in the content of Portland cement. Two of these materials, used most widely are fly ash (pulverized fuel ash) and ground granulated blastfurnace slag (slag). In some cases, there is more fly ash in the mix than Portland cement.

Another very valuable ingredient in concrete, which has slowly become used, is silica fume. I know it is expensive but, without it, very high strengths of concrete cannot be achieved. Here again, there was a delay, Japan being the leader, then the USA, with the UK bringing up the rear.

The present

Returning to admixtures, once again, we were slow in using super-plasticizers. I remember being asked to advise a major British chemical manufacturer on this topic. His final conclusion was that super-plasticizers were valuable but the manufacture of those long and heavy molecules would be expensive, and we should therefore have to rely on imports. To the concrete community this was an expensive solution, or no solution! Superplasticizers were already widely used in the USA and in Germany.

The above may be an unfair criticism: progress is nowhere linear; it was only natural for there to be a pause following the earlier glorious developments. People get tired of change. There is a story about a

19th century politician who said: 'Why should we want change? Things are bad enough as they are.' An opposite view was taken by Benjamin Disraeli – no concrete specialist, he – who said in Edinburgh in 1867: 'Change is inevitable in a progressive country. Change is constant.'

We have to acknowledge that change upsets a working system, even if it is limping; for example, with additional cementitious material, you need another silo for additional materials or a change in concreting operations.

One example of a change in operations is the introduction of self-compacting concrete. This material and technique of placing were developed in Japan. The benefits are three-fold: no noise due to vibrators, which allows concreting at night and at weekends; ability to fill inaccessible spaces within the formwork as well as concreting through reinforcement; and third, avoidance of damage to the skin and nerves of the vibrator operative (the so-called white finger syndrome). The last mentioned problem is about to be tackled by health and safety legislation.

Recently, self-compacting mixes have been extended to include lightweight aggregate.

The intriguing aspect of self-compacting (self-consolidating) concrete is that is was developed in Japan for a totally different reason: in a country where more than one-half of young people are university graduates, operating a vibrator is not an attractive proposition, and the country is opposed to bringing in immigrants from less developed countries. It may be apposite to mention that, for the same reason, it was Japan that pioneered the use of robots in concreting operations. In my opinion, this is a move in the right direction for all of us.

Having said all that, we should not ignore the numerous superb, efficient, economic, and attractive – or even beautiful – concrete structures. They stare at us from the pictures of the Concrete Society Awards and from books such as *Concrete: A Pictorial Celebration*, published in 2004 on the occasion of the centenary of ACI.

The future

It is practical developments that will enable concrete to be used more extensively in the future, just as a wholesale use of robots and computer-based operations has been instrumental in a massive increase in car production and reliability, coupled with a cost reduction. The days of craftsmen producing concrete are numbered, and we do not even have craftsmen, but largely semi-skilled, if not unskilled, labour. These words may be unpalatable but they have to be offered.

There are other practical areas where improvement is necessary: in-line measurement of water on the aggregate; reliable means of ensuring moist curing; development of standardized tests for the properties of self-compacting concrete; verified limits on the impurities in mix water; and a means of measuring shape and texture of aggregate of direct relevance to workability.

Concluding remarks

It is research and development in these and other practical areas that should be financed, and not detailed studies of the shape of expensive fibres for special niche applications, conducted by backroom PhDs. Amusingly, it is a PhD, John Crispo who wrote: 'An Assistant Professor is a PhD who has learned to make a single point into a lecture; an Associate Professor has learned to make a point into a course; a Professor has learned to make a point into a whole career (how true!); a Dean has forgotten the point; and the Principal of the University thinks there never *was* a point.' (I know, I used to be one!)

So, the recent years have been pedestrian. And the future? We have a choice: either it will be problematical, with a decrease in the attractiveness, economic as well as technical, of concrete in construction; or else concrete will get a new lease of life and will be the pre-eminent construction material with a low whole-life-cycle cost. It is the latter alternative that we should strive to achieve.

7.4 THE FUTURE

I had not intended to include in this book a section entitled 'The future' because I lack the breadth of knowledge necessary for a prediction – not only of concrete technology, but also of economics and social aspects. Nor am I influential or powerful enough to will future events. Moreover, every now and again, there occur unexpected but deeply significant events such as a large increase in the price of fossil fuel and gas in the year 2005. Changes in the pattern of international trade and immigration shifts are also bound to affect the use, price, and popularity of concrete.

Nevertheless, I have had to accept the argument put to me that to end my last book, that is, my last view of concrete, by looking backwards would be seen as defeatist. I have, therefore, added this section, which discusses a few developments that I consider imperative as well as practicable. The present situation is not tenable in the longer term. Particularly apposite are the words of Scott Steedman in the Editorial of *Ingenia*, journal of the Royal Academy of Engineering, December 2005: 'construction, quaintly described as a more "traditional" industry'.

What do I see ahead? I am a long way from being a child, and so I see the future 'through a glass, darkly'. I am not attempting to be a soothsayer or to daydream. What I am offering is a wake-up call before our construction sites are run by people who have successfully developed new building materials or perfected some existing ones – but not concrete. In other words, what should we do to retain the primordial position of concrete?

I intend first to look at materials used in concrete and then at procedures in concrete making.

Aggregate surface properties

For environmental as well as fiscal reasons, we are bound to use ever-increasing amounts of recycled concrete aggregate. For this to happen we have to establish a reliable method of assessing the particle properties with respect to their influence on workability. Of course, the same applies to ordinary crushed aggregate, as I have argued for some time. By now, however, I have given up all hope of necessary research being undertaken, because benefits would accrue not to a single enterprise, but would be in the public domain. Thus the quotient cost/benefit would be uneconomically high for a single producer.

I have just admitted to abandoning all hope, but I shall just say once more: the processing of aggregate in terms of true uniformity of grading (superior to finish screening at one, or at most, two screens) and also measurement of texture, which a computer converts to influence on workability in real time, are also necessary in the very near future.

Recycled concrete aggregate

University laboratory studies are needed to establish a quantitative means of determining the shape and texture of aggregate. This will be beneficial for natural aggregates as well as for recycled concrete aggregate. So far, most university studies on recycled concrete aggregate have been concerned with gross physical and mechanical properties of concrete containing this type of aggregate. Such studies are useful, but equally important is the development of economic means of converting demolished concrete at the site of the old structure into usable aggregate at the mixer being used for the new structure. It is the in-between movements and processes that may use energy as well as the space for storing the materials during the intermediate stages. Very little has been investigated in this area.

In Section 5.2, I referred to the use of recycled concrete as aggregate (called recycled concrete aggregate) in new concrete. This is bound to develop very significantly because of two complementary reasons. The first is the decreasing number of sources of rock that can be crushed, coupled with tax on quarrying rock; in some countries tax is levied on all virgin aggregate regardless of its provenance. The second reason is the shortage of disposal sites for demolition material, and specifically for 'old' concrete, again coupled with tax on landfill. This tax has been going up steadily and, in the UK, is now about £15 ($25) per tonne.

The 'paper solution' is: just crush old concrete and use the resulting aggregate to make new concrete. However, as with many construction activities, there is the problem of location or rather locations. This is not always appreciated by electronic engineers or by production engineers who bring all the necessary materials or components to a single location and manufacture the new product under a single roof.

By contrast, let us consider the ideal situation of an old highway that is being demolished and is to be replaced by a new concrete pavement. (There is a caveat that recycled concrete may be vulnerable to freezing and thawing.) Once the old concrete has been broken out, it needs to be crushed, comminuted, and graded to produce aggregate. If this could be done on the demolition site and the new concrete placed in the vicinity, there would be very limited transport charges so that the production of new concrete would be economical.

Let us then first look at crushing the old concrete. I am familiar with only one mini-crusher. This is a rubber-tracked model, weighing 3 tonne, which can be transported on a trailer. The throughput of the crusher is up to 20 tonne per hour, and it can crush old concrete pieces up to 0.5 m by 0.2 m in size, producing aggregate particles nominally 5 mm to 100 mm in size.

So far, so good. However, the new concrete needs fine aggregate as well; that is, a supply of particles smaller than 5 mm. This means stockpiling in the vicinity of a concrete mixer, or else a moving convoy of the crusher, the fine-aggregate supply truck, and the mixer. There is, however, a difficulty over grading. The crushed recycled concrete aggregate needs to be graded and recombined with appropriate particle sizes.

But particle size is not the only criterion for achieving suitable aggregate for the mix. Specifically, the shape of the crushed particles, and their rougher texture, may not be optimal. Water absorption is often high because the pores in the old concrete are now additional pores in the aggregate. For these reasons, the recycled concrete aggregate is often blended with aggregate from other sources. In consequence, we may be faced with the need to have a batching plant to which, and from which, material is transported. Of course, all this represents cost. What we expect in the future are elegant and economic solutions for temporary batching plants.

According to British Standard BS 8500-2, recycled concrete aggregate is allowed to contain up to 5% of masonry. Material with a higher proportion of masonry is called recycled aggregate, and has to be considered separately.

What is needed is a systematic study of mixes using a partial replacement of aggregate by recycled concrete aggregate, combined with logistic arrangements minimizing transport costs and inconvenience. This requires fieldwork and not just a laboratory study of materials.

At present, the leaders in the use of recycled concrete aggregate are the Netherlands, Belgium and Denmark. World-wide, much of the current use of such aggregate is in low-grade applications such as road sub-base or blinding concrete. To contribute to sustainable construction, we should do much better than that and use recycled concrete aggregate in structural concrete.

It is amusing to think that recycled glass is nowadays promoted by the Concrete Technology Unit in Dundee for inclusion in concrete. The

Glass Manufacturing Industry Council in the USA is also looking into this possibility.

My amusement stems from the fact that, some 20 years ago, I was retained by glass manufacturers to deal with the hazard of alkali–silica reaction in glass-reinforced concrete because glass contains alkalis. Actually, glass contains both alkalis (sodium) and reactive silica. However, it is possible to suppress the harmful expansion induced by the alkali–silica reaction through the inclusion of slag or metakaolin in the mix. This is similar to increasing the silica–alkali ratio in the mix, first introduced in 1950.

Anyway, in my view, this is a solution to a problem which need not have been created in the first place. In other words, why use glass as aggregate? The energy required to melt crushed glass is smaller than that of processing sand to make glass. Moreover, crushed glass contains soda ash, which is expensive and is needed to make glass from sand. Glass is too valuable to be 'thrown away' as aggregate: glass should be recycled as glass; whisky has to be put into bottles.

Procedures

Compaction and finishing are activities in which modern techniques that minimize human labour have been developed. Indeed, in many structures, compaction by traditional means, that is, by the use of immersion vibrators, can be avoided. This is achieved by the use of self-compacting (self-consolidating) concrete (see Section 6.1). Robots have been used very successfully, albeit in very few countries, to compact the concrete and also to apply the finish. The noise due to vibration is avoided. A side benefit of avoiding the use of labour in contact with fresh concrete is that chromium in the cement does not lead to dermatitis.

The main benefit of these automatic procedures is that need for labour, largely unskilled, or semi-skilled, is greatly reduced. In Western countries and in Japan, the minimum level of education rises and this results in a reluctance to undertake unskilled tasks, performed often under somewhat unpleasant and certainly uncongenial conditions. Once upon a time, imported, that is immigrant, labour was used, but people from outside Western Europe are increasingly more reluctant to engage in concreting. We often discuss the under-representation of ethnic minorities in various types of employment, but in concreting they are not numerous. Ethnic minority women are conspicuous by their absence, and even women of indigenous background (English) working on a concrete site are a rarity. And yet, there can be absolutely no objection to women concretors.

In the UK, precast concrete has not yet gained the prominence that it has, for example, in Finland. Precast concrete requires less labour on site, especially people involved in 'dirty' work, which some consider unpleasant. Moreover, precast concrete produces structural members of higher quality than routine site work. This applies to compaction (consolidation), finishing, and tolerances on cover to reinforcement. Additional benefits may be in reduced size of the batching plant. I am mentioning all this because I see a reduction in site labour as a desirable objective in the future.

Much development can be achieved: the basic knowledge is available but the devil is in the detail.

Future types of concrete

Let me now turn to a much broader and fundamental issue: the future of concrete as we know it. Earlier in the book (Section 6.1) I referred to low-tech concrete as distinct from high-tech concrete. Low-tech concrete is what is routinely produced by ready-mixed concrete suppliers for a wide range of construction. There exist several basic mixes, differing in the water–cement ratio, in cement content, in workability, and sometimes in the presence of materials such as fly ash or slag, and also admixtures. The scope for fine-tuning of the mix composition is limited, and there is a limited separation, both physical and in terms of decision taking and responsibility, between the specifier and the supplier.

In consequence, improvement in the system is difficult, and what is achieved by Engineer A and Supplier M cannot be readily transferred to a good starting mix by Engineer B and Supplier M, or even by Engineer A and Supplier N. The reason for this is that the materials put into the mixer are dependent on the source of all the aggregates and of cement. This is why, in my opinion, computer-based mix designs work within a limited location only.

The main problem is the variability of cement. To comply with a cement type (according to ASTM) or with a European classification is very easy. In consequence, cements differing substantially in chemical composition or in the reactivity of sulfate, or even in fineness, can all be included in nominally similar mixes supplied by the same ready-mixed concrete supplier. If one complains about this, as I have done more than once, the answer is: the cement is cheap, and you cannot have high-tech material for the price of a low-tech one. Anyway, you have to put up with this situation because there is no existing viable alternative to what we might call the common-or-garden concrete mixes made with whatever cement that the manufacturer chooses to supply.

So we shut up and put up. This situation has existed for a century, but it does not follow that it will continue for ever. I don't know what is a viable alternative to the present-day set-up and, if I knew, I would not disclose it for the paltry sum that the reader has paid for this book!

I admit that the situation is satisfactory for many types of construction but not for all, and not forever. This is why we have, in recent times developed the concept of high-performance concrete. High performance means really fit for the purpose, and everything you buy should be fit for a given purpose. The present situation is not quite 'one size fits all' but it certainly is not a custom-built approach.

High-performance concrete is not the same as high-strength concrete, but high-strength concrete is one example of high performance. Other examples would specify the value of the modulus of elasticity or of creep, possibly of shrinkage or of permeability, reactivity of sulfate, or whatever the designer needs to have in order to achieve a well-performing and adequately durable structure.

We could achieve this, but not with the present supply of cement from any old plant at the cement manufacturer's discretion. The days of small cement plants are all but gone. Nearly all the plants in the Western world and in much of Africa are owned by three mega cement suppliers. This is not a monopoly, but there is no scope for a competitor establishing itself. A competing cement plant would be viable but only if it could sell its cement at a premium price. My opinion is that there is room for such a manufacturer who could create a market. I don't mean a tiny niche market but a small and growing market.

The market will grow as more engineers and architects, and more owners, realize that paying more for a better product is, at least in some cases, worthwhile. A lunch at Pizza Hut is not the same as a birthday party at the Ritz. Indeed, on the food scene, there are hundreds of Asda or Sainsbury stores but there is only one Fortnum and Mason.

The 'high-class' concretes are likely to contain several cementitious materials. Furthermore, producing concretes fit for the purpose usually involves the use of one or more admixtures. Specifically, if compositional requirements (for mechanical or chemical reasons) result in poor workability of the mix, superplasticizers have to be included in the mix. There have been significant improvements in superplasticizers, but we still have to ensure compatibility of every new superplasticizer with the particular Portland cement used. This is so even if the cement originates from the same plant but not from the same consignment as before. The situation is likely to be aggravated by the increased range of kiln fuels used because of the differences in the reactivity of sulfates in the clinker. Laborious testing of mortar using the Marsh cone should be avoided by selling compatible combinations of cement and admixture. Fortnum and Mason can advise on the 'compatibility' of a given caviar with a particular champagne: ready-mixed concrete suppliers should do likewise – at a price, of course. But the price need not be exorbitant if the supplier bears in mind that his business will develop.

So we need a 'Fortnum and Mason', not instead of Asda, but as well. Major producers of cement may feel that enough money can be made out of the cheapest and simplest cement, so they are unwilling to invest in a special plant and elaborate procedures. At the same time, it is not easy for outsiders to penetrate the cement market because a large capital and development outlay would be necessary without profit for some time.

These are not flights of fancy, but a survival plan, which in my view is necessary to keep a place for concrete in construction in the long-term future.

Life-cycle design

This is design that considers in a quantified manner not only the initial cost of construction, but also the cost of operation over a specified period. Cost should include not only financial expenditure but also any inconvenience and interruption in 'enjoyment' of use.

Up to now, life cycle design is not approached in a singled standardized manner. It is easy to take into account the cost of predictable maintenance, but there are other aspects, too. I was involved in the investigation of problems in a high-rise office block in Australia. The owner was concerned not only by the loss of rent when the use of premises was denied to the office tenant because of repairs, but was also conscious of the negative effect of external work on potential tenants.

In a different sphere, in the 1970s, blocks of flats built in concrete containing high-alumina cement were difficult, if not impossible, to sell. They were subject to blight, even if the particular building was not condemned on safety grounds.

The most advanced life-cycle design is that of important bridges; otherwise, the design life is usually limited to considering isolated single parameters, one at a time, such as chloride penetration leading to corrosion of reinforcement. Highway pavements are designed for a certain number of axle loads. However, all this is short of aircraft design, both in terms of fatigue of metal in the body (of the plane) and reliability of engines.

Hopefully significant advances in life cycle design of concrete sections will come before long.

Future knowledge required

We have quite a good understanding of transport of fluids through concrete: we should now translate it into design of mixes with the desired relevant properties. So-called waterproofing admixtures are often used successfully, but knowledge of these admixtures is not available in a reliable and independent form. The specific adverse effects of a very low permeability of concrete with a very low w/c in the case of fire need further study so as to resolve the problem of bursting of concrete on expansion of steam bubbles.

In some court cases, the plaintiffs try to find out areas of scientific lack of clarity and uncertainty, if not outright confusion. This they then use as the battleground where their witnesses, if not entirely scrupulous, hold forth with an air of authority. If this cannot be nipped in the bud, for example by a motion *in limine* (see Section 2.3), the jury is likely to 'buy' a totally erroneous proposition. Because scientists are better able to expound on scientific principles (true or false) than engineers, there is a real danger that an engineering expert becomes cornered and, in cricket parlance, 'retires hurt'.

It is, therefore, important to establish rationally, lacunae in our knowledge, and to show the areas that need reliable clarification before inferences can be presented in court. Section 4.3 dealing with the 'Confused world of sulfate attack' is a prime example. Likewise, Sections 4.2 and 5.1 also illustrate the limits of our knowledge and the consequences thereof.

It would be impossible not to acknowledge the large contribution to better concrete in practice made by Mohan Malhotra. I referred in Section 6.1 to his personal research and development of concrete with a very high content of fly ash. However, he has done more than that: since 1992, he has organized eight conferences on advances in concrete technology. These are useful, but some of the topics are inevitably academic or apply to concretes used in highly specialized situations: these are micro-niche applications. My vision of the future is concerned with large-scale use of concrete in important structures, such as bridges and buildings, and also with housing and highways. My discussion of the future has to be seen in those terms. Lack of studies in that area may be the consequence of the fact that frequently the academics attending conferences have a narrow and exclusively academic or scientific background. Teachers of engineers need to include many practitioners, and in the UK, and many Commonwealth countries, this is not the case: the possession of a PhD is often seen as of paramount importance in selecting a staff member. To remedy this situation is not easy.

I may have painted a sombre and even pessimistic picture of the future of concrete. Nevertheless, my motives are positive; I want concrete to blossom (figuratively speaking and not by efflorescence) and I wish I could say with full confidence: we shall succeed! Or shall we?

7.5   A FAREWELL TO CONCRETE

In many people's eyes, my name is associated with concrete, so much so that, if I lived in Wales, I would probably be known as 'Neville the Concrete'. The title of the Japanese translation of the Fourth Edition of *Properties of Concrete* is written in Japanese characters to represent the English phonemes of 'Neville's Concrete Bible'.

But the notion of the exclusive association of my name with concrete is not correct. Apart from working on a geothermal power project in New Zealand, which included a wide range of civil engineering works, and on hydroelectric power design, I spent many years teaching at universities, the subjects ranging from structural analysis to hydraulics and to statistical methods, as well as surveying and geodesy. Some of my university career involved being the Dean of the Faculty of Engineering, Dean of Graduate Studies, and, for nine years, service as Principal and Vice-Chancellor (President) of the University of Dundee: this university comprised Faculties of Arts, Science, Law, Medicine and Dentistry as well as Engineering.

Nevertheless, my first book was on concrete, my very extensive consulting work was on concrete, and I never gave up my interest in concrete and concrete structures. More than that: I suborned my wife to join me in my concrete 'enterprise', starting when we were engaged to be married and continuing to the production of this book 54 years later. It's a long time, she says. So it would be appropriate to say, on her behalf as well as my own, goodbye concrete – it was good to know you, but enough is enough.

Finally, my concrete work gave me a chance to make numerous good friends, literally the world over. Perhaps that has been the greatest 'side effect' of my life 'in concrete'. I echo Mathew Arnold:

> Is it so small a thing ...
> To have loved, to have thought, to have done;
> To have advanced true friends, and beat down baffling foes?

Appendix

DETAILS OF ORIGINAL PUBLICATIONS

Most sections of this book were first published as papers and articles; relevant discussions are included. Their details are listed below.

Chapter 2 Understanding the water–cement ratio

Section 2.1 Aïtcin, P.-C. and Neville, A., How the water–cement ratio affects concrete strength, *Concrete International*, 25, Aug., 2003, pp. 51–58.

Discussion, *Concrete International*, 26, April, 2004, p. 9.

Section 2.2 Neville, A., How closely can we determine the water–cement ratio of hardened concrete?, *Materials and Structures*, 36, June, 2003, pp. 311–318.

Chapter 3 High-alumina cement

Section 3.1 Neville, A., Draft standard for high-alumina cement: should it tell us how to make concrete?, *Concrete*, 37, July, 2003, pp. 44–45.

Section 3.2 Neville, A., Revised guidance on structural use of high-alumina cement, *Concrete*, 38, Sept., 2004, pp. 60–62.

Section 3.3 Neville, A., Should high-alumina cement be re-introduced into design codes?, *The Structural Engineer*, 81, Dec., 2003, pp. 35–40.

Discussion, *The Structural Engineer*, 82, 10, May, 2004, pp. 38–39.

Discussion from India, Correspondence, *The Structural Engineer*, 82, July, 2004, p. 36.

Chapter 4 Durability issues

Section 4.1   Neville, A., Can we determine the age of cracks by measuring carbonation? Part 1, *Concrete International*, 25, Dec., 2003, pp. 76–79.

Section 4.2   Neville, A., Can we determine the age of cracks by measuring carbonation? Part 2, *Concrete International*, 26, Jan., 2004, pp. 88–91.

Section 4.3   Neville, A., The confused world of sulfate attack on concrete, *Cement and Concrete Research*, 34, Aug., 2004, pp. 1275–1296.

Section 4.4   Neville, A. and Tobin, R.E., Sulfate in the soil and concrete foundations, *Concrete Construction*, 50, 3, March, 2005, pp. 74–77.

Section 4.5   Neville, A., Background to minimising alkali–silica reaction in concrete, *The Structural Engineer*, 83, 2005, pp. 18–20.

Chapter 5 Behaviour in service

Section 5.1   Neville, A., Which way do cracks run?, *Journal of ASTM International*, 2, 8, Sept., 2005.

Section 5.2   Neville, A., Some aspects of sustainability, Guest Editorial, *Prestressed/Precast Concrete Journal*, 51, 1, Jan./Feb., 2006, pp. 72–75.

Section 5.3   Neville, A., Requirements for residential slabs on grade: Part I – The ACI approach, Part II – Uniform and International Codes, *Residential Concrete Magazine*, June, 2005, pp. 33–42.

Section 5.4   Neville, A., Requirements for residential slabs on grade: Part III. Who selects the mix for residential slabs on grade?, *Residential Concrete Magazine*, Sept., 2005, pp. 41–43.

Chapter 6 General issues

Section 6.1 Neville, A., Concrete: from mix selection to the finished structure – problems en route, *Prestressed/Precast Concrete Journal*, 49, 6, Nov./Dec., 2004, pp. 70–78.

Section 6.2 Neville, A., Workmanship and design, *Concrete*, 39, 11, 2005, p. 89.

Section 6.3 Neville, A., Relevance of litigation to the structural engineer, *The Structural Engineer*, 82, Oct., 2004, pp. 10–14.

Section 6.4 Neville, A., Violation of codes, Letter, *Concrete International*, 26, March, 2004, p. 17.

Section 6.5 Neville, A., Gender in concrete, *Concrete*, 36, May, 2002, p. 52.

Chapter 7 An overview

Section 7.1 Neville, A., Concrete: 40 years of progress??, *Concrete*, 38, Feb., 2004, pp. 52–54.

Section 7.2 Neville, A., Looking back on concrete in the last century, *Concrete*, 39, March, 2005, pp. 37–38.

Section 7.3 Neville, A., Concrete past and present, *Concrete*, 39, April, 2005, pp. 50–51.

Index

Index

308